# Creation, Evolution, And Catholicism

## A Discussion for Those Who Believe

By
Thomas L. McFadden

Institute for Science and Catholicism

In order for this book to discuss the scientifically and religiously controversial subjects that it has it was necessary to identify and quote from published sources on each side of those subjects to criticize or comment on them, to support the author's thesis, to illustrate by examples, or for other fair uses in accordance with section 107 of the U.S. Copyright Act which permits "fair use" of copyrighted material in certain circumstances without prior permission. Guidance for interpretation and application of section 107 provided by the Copyright Office online was followed. This book is an educational outreach project of the Institute for Science and Catholicism (ISC) which is solely responsible for its content. ISC acknowledges in particular **Creation Ministries International** whose website provided the author with an enormous reservoir of science education and many references in this book.

ISC is a non-stock Virginia corporation organized for charitable, religious, and educational purposes within the meaning of section 501(c) (3) of the Internal Revenue Code. ISC is promoting the renewal of a Catholic theology of creation and a new science/faith synthesis based on sound scientific data and a serious approach to the Holy Scriptures in accordance with longstanding Church Tradition.

Email: SCIENCEandCATHOLICISM@GMAIL.COM.

On the web: ScienceAndCatholicism.org

On Facebook: Institute for Science and Catholicism

**Library Cataloging Data**

*Creation, Evolution, and Catholicism: A Discussion for Those Who Believe* by Thomas L. McFadden

282 pages, 6" x 9" x 0.69" & 0.93 lb.

BISAC: SCI027000 SCIENCE / Life Science / Evolution

BISAC: SCI008000 SCIENCE / Life Science / Biology

BISAC: SCI075000 SCIENCE / Philosophy & Social Aspects

BISAC: SCI034000 SCIENCE / History

BISAC: REL106000 RELIGION / Religion & Science

BISAC: PHI010000 PHILOSOPHY / Movements / Humanism

**ISBN-13:978-1530654765**

**ISBN-10:1530654769**

**Library of Congress Control Number: 2016906195**

**CreateSpace Independent Publishing Platform,**

**North Charleston, SC**

# Contents

"When men cease to believe in God, they do not then believe in nothing, they believe in anything."

**G. K. Chesterton (1874-1936), writer, philosopher**

"Theistic evolutionists are deluded."

**Richard Dawkins, best-selling atheist author**

"To say that animals evolved into man is like saying that Carrara marble evolved into Michelangelo's *David*."

**Tom Wolfe, best-selling novelist**

"Ever since the creation of the world His eternal power and divine nature, invisible though they are, have been understood and seen through the things He has made. So they are without excuse."

***Romans* 1:20**

"I charge you in the presence of God and of Christ Jesus who is to judge the living and the dead, and by His appearing and His kingdom: preach the word, be urgent in season and out of season, convince, rebuke and exhort, be unfailing in patience and in teaching. For the time is coming when people will not endure sound teaching, but having itching ears they will accumulate for themselves teachers to suit their own likings, and will turn away from listening to the truth and wander into myths. As for you, always be steady, endure suffering, do the work of an evangelist, fulfill your ministry."

**2 *Timothy* 4:1-5**

# Preface

The accelerating loss of faith by Catholic youth is reported in survey after survey. An estimated 15 million Americans have left the Catholic Church since 2000. Today roughly 50% of American adults under 30 do not believe in a personal God. That has social and political consequences as well as spiritual consequences. Once a conscious choice is made to reject God's truth, an individual begins to hate the people of God.

Molecules-to-man evolution—the origin of man and of all living things (with or without divine assistance) through hundreds of millions of years of the same kinds of material processes going on now—is taught to most students as a scientific fact in public and Catholic schools and by the culture at-large. This book will demonstrate that this evolutionary indoctrination plays a significant role in the accelerating loss of Faith among youth.

Years ago I became aware of the skepticism among Catholic teens while teaching CCD when they asked me "You don't really believe in Adam and Eve, do you?" I learned from them that their unbelief in supernatural doctrines derived from the Bible was because of their belief in evolution. They realized there was a conflict between the Book of *Genesis* and the "science" they were taught in school; their school teachers were more effective than their religious educators and had so much more of the students' time to make their case. At that point, nothing in my educational background or engineering career had acquainted me with the subject of evolution. When I did research on evolution I discovered that many people believe that it is an undeniable scientific fact that our origins were in cosmic and biological evolution. Many who accept that belief likewise logically conclude that no God is necessary and that *fiat* creation by a loving, personal and interventionist God is unthinkable. World-famous atheist Richard Dawkins said it best: "Darwin made it possible to be an intellectually fulfilled atheist." I found that evolution is the basic dogma of a non-theistic religion called Humanism that considers Christianity harmful to the common good and teaches that it must be replaced. That belief and the public policies advocated by Humanists were clearly expressed in the creed-like *Humanist Manifesto I* of 1933 and *Humanist Manifesto II* of 1973.

The public policies advocated by the non-theistic religion have become the anti-Christian normal at all levels of government and other Humanist-controlled institutions such as education, entertainment and news media. The shocking, hateful post-election vitriol poured out on Trump and those who voted for him by the Humanist Left is beyond just "politics." It is Humanist religious rage that will end badly for Trump and us when the Abortion Party regains the Presidency. To sustain its cultural dominance Humanism requires a constant supply of practical atheists. It cultivates our children in the schools and colleges. When indoctrinated with the dogma of naturalism (evolution) children are conditioned to become skeptical and often hostile to any supernatural religion and its moral values.

To save our children and our culture Catholic clergy and laity must cooperate at the parish level to refute the bogus scientism which is the cornerstone dogma of Humanism and to teach the doctrine of creation that is the foundation of the Faith.

If that plan were to be followed it would kick start the New Evangelization of which we hear where it is needed most-- within our own Catholic community. In so doing I assert that Catholics will recover the outgoing confidence needed to keep our youth Catholic and become more capable of engaging our culture.

Thomas L. McFadden, Sr.

# Chapter 1-The Crisis of Faith

Pope St. John Paul II once said that every generation, with its own mentality and characteristics, is like a new continent to be won for Christ. In the United States, that battle is being lost. As will be shown, Catholic youth are leaving the Faith at an accelerating rate. Adults who are committed and practicing Catholic parents of young children might not even think about the extent of the problem if it is has not happened in their family. How would parents who go to Church and their priests, surrounded by like-minded people, know how many youth went "missing" intellectually so soon after receiving their confirmation?

**The Overall Catholic Loss**

The Pew Research Center's Religion and Public Life Project, Religious Landscape Survey, *Religious Affiliation: Diverse and Dynamic* February 2008 found that belief and practice among adult Catholics in the U.S., particularly the younger ones, is in a steep decline:

> While those Americans who are unaffiliated with any particular religion have seen the greatest growth in numbers as a result of changes in affiliation, Catholicism has experienced the greatest net losses as a result of affiliation changes. While nearly one-in-three Americans (31%) were raised in the Catholic faith, today fewer than one-in-four (24%) describe themselves as Catholic.

The Survey report also cites the Catholic exodus as an example of the dynamism of the American religious scene:

> Other surveys – such as the General Social Surveys, conducted by the National Opinion Research Center at the University of Chicago since 1972 – find that the Catholic share of the U.S. adult population has held fairly steady in recent decades, at around 25%. What this apparent

stability obscures, however, is the large number of people who have left the Catholic Church. Approximately one-third of the survey respondents who say they were raised Catholic no longer describe themselves as Catholic. This means that roughly 10% of all Americans are former Catholics.

## Falling Away

As noted above, the research shows that 1 in 3 Americans were raised Catholic but only 1 in 4 still self-identify as a Catholic. Year after year, survey after survey also confirms that the majority of those who continue to self-identify as Catholics in the U.S. do not accept various Church teachings and that the younger the survey respondent the more extensive the non-acceptance. For example, among the numerous surveys are ones commissioned by the United States Conference of Catholic Bishops, such as *Sacraments Today: Belief and Practice among U.S. Catholics,* completed in February 2008 among people who still consider themselves to be Catholics. The lack of real belief and practice among those who at least self-identify as Catholics is reflected in Mass attendance, which Pew Research found is about 1 in 5.

As an example of anecdotal support for the Pew findings regarding Mass attendance, consider this February 2015 observation by blogger Fr. Joseph Illo (frilloblog.com):

> The Catholic Church in San Francisco is but a shadow of what she once was, and I suppose that is true in most of our big cities…'But father, I've tried to bring my children back to Mass!' many lament. True enough: it's downright difficult to serve people who don't want our services. The 50 Catholic parishes in San Francisco offer hundreds of "services" every Sunday, and most are mostly empty.

The relative emptiness of some big city parishes is certainly due, in part, to population shifts of families from the cities to the suburbs, both because of the cost and the quality of life for children. But on the other hand, demographic data in many cities indicate that young, upwardly mobile professionals ("yuppies") are moving in. Among those are a proportional share who were raised Catholic. They are the well-educated potential "new blood." But too many are lost along the way.

### Church Officials and Broken-Hearted Parents Know It

In the March 2016 issue of *Faithful Insight* the Bishop of Phoenix showed that Bishops have noticed what broken-hearted parents have been observing for decades, namely, that Catholics are leaving in droves. According to the Most Reverend Thomas Olmstead since 2000

> 14 million Catholics have left the faith, parish religious education of children has dropped by 24 percent, Catholic school attendance has dropped by 19 percent, infant baptism has dropped by 28 percent, adult baptism has dropped by 31 percent and sacramental Catholic marriages have dropped by 41 percent.

Bishop Olmstead called those losses "staggering." The Bishop went on to say that

> One of the key reasons the Church is faltering under the attacks of Satan is that many Catholic men have not been willing to 'step into the breach'-to fill this gap that lies open and vulnerable to further attack.
>
> A large number have left the faith, and many who remain 'Catholic' practice the faith timidly and are only minimally committed to passing the faith on to their children.

Had the Bishop not telegraphed that he was blaming the laity by mention of "their children" he could have been describing a significant segment of the episcopacy and diocesan clergy. The Church that Our Lord founded is a hierarchical organization with leaders over the laity appointed from the ranks of the clergy by the bishops. If McDonalds lost as customers 14 million children of its current customers, would McDonald's stockholders blame the current customers? They might blame McDonald's corporate management. Corporate management would realize it has to do a lot of things differently. The objective of this book is to promote one change that will help us keep our youth Catholic.

**2014 Update: Catholicism in America is Not Growing**

The Pew Research Center, Religion & Public Life published a report May 12, 2015 called "America's Changing Religious Landscape" based on surveys taken in the summer of 2014. The data show that between 2007 and 2014 "the Christian share of the population fell from 78.4% to 70.6% driven mainly by declines among mainline Protestants and Catholics." Keep in mind that in these surveys Pew classifies respondents based on how they self-identify so the statistics include the committed and the nominal. The "Catholic" part of the population dropped from 23.9% in 2007 to 20.8% in 2014. The "unaffiliated" share of the population that includes atheists, agnostics and former Christians increased to 22.8%. And, reflecting the fact that Catholic youth are becoming unaffiliated, the median age of Catholic adults in 2014 was 49. It was 45 in 2007. On November 4, 2015 Pew published additional finding based on the questions asked during the surveys taken in the summer of 2014. Pew found that women are more likely than men to be certain about God's existence (69% vs. 57%), as are less well-educated Americans compared with college graduates (66% vs. 55%). While 70% of persons over 65 are certain that God or some Supreme Being exists, only 51% of adults under 30 are. Among persons who self-identified as

Catholics, only 64% expressed an absolutely certain belief in God in 2014, down from 72% in 2007.

### Non-Religion is the Fastest Growing "Religion"

It is not just the faith of Catholics in America that has been lost. Pew Research surveys found evidence of a gradual decline in religious commitment in the U.S. public as a whole. For example, a July 2013 Pew Survey, "Growth of the Nonreligious," reports that "there has been a modest uptick over the past decade in the share of U.S. adults who say they seldom or never attend religious services." The number of Americans who do not identify with any religion at all has also grown in recent years; indeed, about one-fifth of the public overall – and a third of adults under age 30 – are religiously unaffiliated as of 2012. Fully a third of U.S. adults say they do not consider themselves a "religious person." And two-thirds of Americans – affiliated and unaffiliated alike – say religion is losing its influence in Americans' lives. If this was ever a predominately Christian country, something has turned its youth into agnostics, deists, or atheists. Catholic and Evangelical pastors lament how reaching Americans is getting more difficult all the time. The soil where the seed of the Gospel is sown continues to grow harder, overgrown with weeds of doubt and confusion, with a resultant diminishing of the Christian faith across the board.

### Specifically Catholic Studies

*Our Sunday Visitor Weekly* published on August 27, 2016 an article titled "Young people are leaving the faith. Here's why: Many youths and young adults who have left the Church point to their belief that there is a disconnect between science and religion" The article was based on two national studies done by the Center for Applied Research in the Apostolate (CARA).

> The interviews with youth and young adults who had left the Catholic Faith revealed that the typical age for this

decision to leave was made at 13. Nearly two-thirds of those surveyed, 63 percent said they stopped being Catholic between the ages of 10 and 17. Another 23 percent say they left the Faith before the age of 10. Those who leave are just as likely to be male as they are female, and their demographics generally mirror those of all young Catholics their age. So why are they leaving?

According to the article they are leaving because of "science." The "disconnect between science and religion" means that the materialism/naturalism explanation of origins taught in school destroys belief in the Bible and the supernaturalism upon which Catholicism depends. This book is intended to cure that problem.

**A Common Denominator**

The one thing that the majority of Catholic youth have in common with their fellow Americans is that they were taught, from the earliest days through high school, the materialistic evolutionary theory of origins as a scientific fact. They were taught that in public schools and in many (if not most) Catholic schools. That teaching is reinforced throughout our culture by science and nature- themed programs produced for the Public Broadcasting System (PBS), History Channel, Smithsonian Channel and all public educational sites such as libraries, natural history museums and National Parks. For decades, Hollywood has been presenting images of 'Martians' and 'space creatures' through movies such as *Independence Day* and *Alien*. Behind this is the idea of evolution: if life evolved on earth, then why couldn't it have evolved on other planets?

The net result is that the majority of students understand evolutionary origins to be a proved fact. A Catholic reading this who also believes evolution to be a proved fact, may be wondering what this writer's problem is with that.

Understandably so, because the majority of American Catholics accept that evolutionary theories of origins are factual. According to a study published in December 2013 by the Pew Research Center,

> Six-in-ten Americans (60%) say that "humans and other living things have evolved over time," while a third (33%) reject the idea of evolution, saying that "humans and other living things have existed in their present form since the beginning of time." The share of the general public that says that humans have evolved over time is about the same as it was in 2009, when Pew Research last asked the question.
>
> About half of those who express a belief in human evolution take the view that evolution is "due to natural processes such as natural selection" (32% of the American public overall). But many Americans believe that God, or a Supreme Being, played a role in the process of evolution. Indeed, roughly a quarter of adults (24%) say that "a supreme being guided the evolution of living things for the purpose of creating humans and other life in the form it exists today."

The report notes that 68% of white, non-Hispanic Catholics believe that humans evolved over time and just 26% believe that humans existed in present form since the beginning. The only groups with a higher belief in human evolution than white, non-Hispanic Catholics are the unaffiliated (76%) and mainline Protestants (78%). Among white Evangelical Christians, 64% believe that humans were created as they are now, just as the Fathers, Doctors, Councils and Popes have taught.

A follow up survey by Pew in 2014 found that belief in evolution continues to trend upward when compared to the results published in 2013 that were the basis for the discussion above. The year to year survey result is shown in tabular form. The surveys were taken in April 2013 and August 2014. No breakdown by religion was published for the 2014 survey

| | | April 2013 | August 2014 |
|---|---|---|---|
| Humans have evolved | | 60% | 65% |
| | Due entirely to natural process | 32% | 35% |
| | Supreme Being guided evolution | 24% | 24% |
| | Evolved but don't know how | 4% | 5% |
| Humans have existed in present form since beginning | | 33% | 31% |
| Don't Know | | 7% | 4% |

**Theistic Evolution**

As shown in the table above, 24 out of 100 American adults find a "third way" between the scientific consensus and the text of the Bible. While accepting that "something" turned into "everything" over billions of years, as taught to them in school, they overlay it with the belief that some Supreme Being guided evolution. The combination of belief in evolution as a proved scientific fact but then overlaid with belief in guidance by God at one or more points in, or prior to, a supposed multi-billion year process defines the theistic evolution theory of origins. Among theistic evolutionists there can be great variation of belief about what those supernatural interventions were and when they happened. Some Catholics who hold that combination have been taught philosophical proofs for the existence of God and have been told that evolution was the playing out of secondary causes flowing according to Divine Providence from the original "whatever it was" created from nothing "whenever." For them it bridges the gap. It "works" for them. Since they have reached mature adulthood and feel their Faith is fully intact, it is practically impossible to convince them that belief in evolution is causing others to lose their Faith. Many Catholics simply "tune out" to objections to evolution, such as the lack of scientific evidence, and other rational arguments. Others accuse fiat creationists of being an embarrassment to the Church by being so "backward."

Belief in evolution often leads to a situation where Catholics lose respect for Catholics who dissent from Darwinian orthodoxy. Often this antagonism is associated with little understanding of the ideological bias of "evolutionary science," and how the "settled science" taught in school is so different from the science discussed in peer-reviewed professional journals.

Relying on the maxim that "there can be no conflict between true science and true religion because God is author of both," many Catholic intellectuals, lay and clerical, sincerely believe that theistic evolution blends faith with scientific credibility. But the school kids are not "buying" it and the refusal to acknowledge that they aren't buying is self-inflicted blindness.

Richard Dawkins, world-famous evolutionary biologist and author of God-denying books such as *The God Delusion* and *The Blind Watchmaker* ridicules theistic evolutionists: "Theistic evolutionists are deluded." Watch this 1-minute video as Dawkins explains why. It is a powerful testimony from an atheist who understands the incompatibility of evolution and Christianity.
https://www.youtube.com/watch?v=BAbpfn9QgGA

# Chapter 2-When Do They Stop Believing?

When Pew interviewed Catholics who had been raised Catholic and who now self-identified as unaffiliated it found that 48% of them lost their faith by age 18 and another 30% lost it by age 23. The specifically-Catholic studies by CARA cited in the last chapter found the typical age at which Catholic children are lost is 13. The significance of this statistic cannot be overstated. At the very least it hints at the importance of counter-Catholic influences in formal schooling, which is the principal occupation of most persons younger than 18 and of many younger than 23. Many still-believing adult Catholics who have had similar schooling are reluctant to accept that something in particular at school could be a primary reason why others have lost their faith.

### Why They Say They Left

Why did these Catholics, 78% of whom left by age 23, say they left? Respondents in the Pew Survey were free to give more than one reason. According to the survey

> When asked to say whether or not each of a number of specific items was a reason for leaving Catholicism, most former Catholics say they gradually drifted away from Catholicism. Nearly three-quarters of former Catholics who are now unaffiliated (71%) say this, as do more than half of those who have left Catholicism for Protestantism
>
> Majorities of former Catholics who are now unaffiliated also cite having stopped believing in Catholicism's teachings overall (65%) or dissatisfaction with Catholic teachings about abortion and homosexuality (56%), and almost half (48%) cite dissatisfaction with church teachings about birth control, as reasons for leaving.

Because most of the respondents lost their faith by age 23, one suspects that the secondary reasons survey respondents gave, such as disagreement about Church teaching on abortion, birth control and homosexuality, were just thrown in to reflect the politically correct positions taught by the American secular culture that has abolished notions of personal sin rather than personal challenges the respondents actually experienced in their youth. In other words, it has not been shown that Catholics drop out primarily because of one teaching or cluster of teachings. Most (71%) said they "just gradually drifted away" and 65% "stopped believing in Catholicism's teachings overall." Philosopher D.Q. McInerny has made the connection between the loss of the sense of sin and loss of belief in God:

> But what is this reluctance to accept the reality of sin but a form of what might be called soft atheism? People who are confused about the reality of sin are very likely confused at a much deeper level—about the reality of the God who is offended by sin.

When a Catholic "gradually drifts away" and "stops believing in Catholicism's overall teaching" did he stop believing in Catholicism's overall teaching and then stop believing in Catholicism's God? Or did he stop believing in God and just gradually drift away?

Based on a follow-up survey in 2015, Pew Research published on August 24, 2016 a report called "Why America's 'nones' left religion behind" that said:

> About half of current religious "nones" who were raised in a religion (49%) indicate that a lack of belief led them to move away from religion. This includes many respondents who mention "science" as the reason they do not believe in religious teachings, including one who said "I'm a scientist now, and I don't believe in miracles."

> Others reference "common sense," "logic" or a "lack of evidence" – or simply say they do not believe in God.

Those responses are similar to those in the CARA studies. It would be comforting to today's parents if the respondents in the earlier Pew surveys who said they were raised Catholic but no longer describe themselves as Catholic were all the children of lax Catholics who made no effort. But there were indicators that those parents made an effort. For example, the survey found that 71% of those who are still Catholic attended religious education classes regularly. Yet 68% of those who are now unaffiliated said they did also. The same percentage of "still Catholic" and "now unaffiliated" said they participated in Catholic teen youth groups. Of the "still Catholic," 25% attended a Catholic high school, but so did 20% of the "now unaffiliated."

### Lack of Necessity May Lead To Unbelief

An article called "Scientists discover that atheists might not exist, and that's not a joke" on the website science20.com reviewed some studies from science journals. The opening lines were:

> Metaphysical thought processes are more deeply wired than hitherto suspected. While militant atheists like Richard Dawkins may be convinced God doesn't exist, God, if he is around, may be amused to find that atheists might not exist. Cognitive scientists are becoming increasingly aware that a metaphysical outlook may be so deeply ingrained in human thought processes that it cannot be expunged.

Those studies support anthropologists who have found that throughout all of history people of all civilizations have in some way believed in a Supreme Being. In other words, we are "wired" or "programmed" to believe in God. So why is doubt and indifference to God the fastest growing religion in America?

## Naturalism: the Root of the Problem

In 1884 Pope Leo XIII identified the problem of naturalism and evolutionism. In his 1907 encyclical, "On the Doctrine of the Modernists" Pope St. Pius X described Modernism as "the synthesis of all heresies."

> First of all [the Modernists] lay down the general principle that in a living religion everything is subject to change, and must in fact be changed. In this way they pass to what is practically their principal doctrine, namely, evolution. To the laws of evolution everything is subject.

Phillip Campbell (*Unam Sanctam Catholicam*) explained that

> the reason Modernism is the synthesis of all heresies is not because it professes all heresies formally, but because of its incorporation of the principle of evolution as applied to truth. Darwin had presented the world with a model of reality which stressed becoming over being; in fact, there really was no "being" in the Aristotelian-Thomist sense. Every "being" was merely a moment in the history of becoming. That being the case, it was only so long before this concept was applied to revealed truth and even God Himself, and thus the Modernist theological school proposed that dogma can in fact evolve, not just in expression but in substance, which is a logical consequent of affirming the evolution of material substances. This is the sense in which Modernism is a synthesis of all heresies: because truth itself is subject to change, dogma becomes a potent medium for the impression of *any* teaching. Once the evolution of dogma is admitted, every heresy is present in potency.

Catholic philosophy professor Dennis Q. McInerny explained the connection between Modernism and evolution this way:

Naturalism is a doctrine which simply denies the reality of the whole supernatural order; it goes hand in hand with materialism. One of the major outgrowths of naturalism was the theory of evolution, which was firmly set in place with the publication of Charles Darwin's *The Origin of Species* in 1859. More than just a scientific theory, evolutionism soon became for many a general philosophy of life, and the Modernists were much taken by it, welcoming it as a sterling manifestation of 'progressive' thought.

### A Substitute Religion

Catholics who told the Pew researchers that "they just gradually drifted away" probably didn't make a conscious decision and suddenly "stopped believing in Catholicism's teachings overall." The decision was made at a deeper level of consciousness. Never in those years of school did the textbooks and teachers need to explicitly say that God does not exist. It was sufficient to show He was unnecessary because "science" explained to them their origin, existence, and physical reality better than the Catholicism they had learned. The worldview based on evolution is a substitute religion. And that religion with no God provides moral autonomy so "if it feels good, do it." Evolutionary biologist and best-selling atheist author Richard Dawkins explained that Darwinism makes theistic belief both implausible and unnecessary: "Darwin made it possible to be an intellectually fulfilled atheist." Dawkins claims to be one.

A school child will believe the story of evolutionary cosmic and biological origins because it is repeated by the authorities. Usually no trusted adult will teach him differently. If he retains some sort of belief in God, his belief can become skeptical. If God exists, then billions of years ago He put some physical laws in place, and has since practiced non-intervention in the natural behavior of the universe. The religious education he receives with its “miracle stories” of a knowable, loving God clashes with cold, hard “science” that credits the formation of life and the universe to only natural processes. The homilies of priests, Church documents, CCD instruction materials, etc. invariably refer to or quote Bible texts with the assumption that those texts are taken at face value, i.e., “as gospel”? That’s no longer a safe assumption to which priests and others have not adjusted. What about the teenagers wrestling with the scientific materialism drummed into them at school? To such teens, if they are even paying attention, instruction depending on “Bible stories” will seem facile. Put simply, church and CCD attendance does not a Christian make.

The evidence that children are leaving in droves because instruction at school in bogus cosmic and biological evolution “science” creates a perceived conflict with religion has been “stacked and catalogued.” Yet, priests and parish Directors of Religious Education just keep doing the same old, same old that has failed for the last 50 years.

Catholic apologetics needs to embrace the 21st Century natural science that refutes those bogus 19th Century theories that have undermined Catholic confidence. The creation doctrines that the spokesmen for the Church seem to have forgotten or misplaced must be taught again. When Catholics recover those doctrines they will be capable of refuting ideas now asserted by Humanists without opposition. The Humanist worldview and the confidence Humanists exude as they steamroll Christians, is based on two affirmations of their faith, evolutionary cosmology and biology, that they have taught the majority of Catholics to accept, at least implicitly. In the following chapters evolutionary cosmology and biology will be considered. That will be followed in chapter 8 by analysis of the last Magisterial teaching that was issued regarding evolution, namely, the encyclical *Humani Generis*.

# Chapter 3-Evolutionary Cosmology

An illustration of theistic evolution--that combination of naturalism overlaid with supernaturalism--was published in a weekly Catholic newspaper revered by its subscribers for its fidelity to the Magisterium. The belief in that combination is proof that it "works" for some Catholics because it is unquestionably true that the man who wrote the illustrative article is a faithful, orthodox, and possibly saintly man who obviously loves the Church. The only purpose of highlighting his opinion is to show how what one learns in school encourages life-long belief in cosmic and biological evolution that when overlaid with God's guidance leads to something unrecognizable as either scientific or Catholic. In his "First Teachers" column of January 22, 2015, in *The Wanderer*, James K. Fitzpatrick wrote:

> The Big Bang, which today we hold to be the origin of the world, does not contradict the intervention of the divine Creator, but, rather, requires it. Evolution in nature is not inconsistent with the notion of creation, because evolution requires the creation of beings that evolve.

Is that what the Catholic Church teaches is the origin of the world? It isn't. In the 1940s one of Britain's best-known astronomers, Sir Fred Hoyle, proposed the "steady state" theory, a belief that the universe had no beginning or end, but always existed and would continue to exist. Humanists, as Hoyle was, rejected any theory that seemed to teach a beginning for the universe because that would point to a Beginner. That bias was so strong that they promoted a theory that violates the fundamental Law of Conservation of Mass/Energy, which states that mass/energy in the universe can neither be created nor destroyed. Hoyle's theory requires a continual spontaneous stream of hydrogen atoms from nothing. Because the evidence of the rapid expansion of the universe exceeded the predictions of Hoyle's theory, and because of their reluctance to accept a theory dependent on violation of that conservation law, many astronomers began to postulate that an explosion of highly dense matter was the beginning of all space and time. In his 1950 BBC radio series, *The Nature of the Universe*, Hoyle mockingly called this idea the "big bang," considering it preposterous. Yet the theory—and the derisive term—have become mainstream, not only in astronomy but in society as well.

### What is that Theory?

It is an attempt to explain the universe as a purely material event. The Big Bang Theory proposes that, at some moment, billions of years ago, all of space was contained in a dense and hot single point from which the universe has been expanding and cooling ever since. There is no scientific consensus regarding the source and cause of that dense mass and the intense heat energy in that imagined single point. In 1980, another hypothetical was added to the theoretical story - the inflation theory. This theory attempts to explain the expansion from the "single point" to the enormous size of the universe. It includes features contrary to known physical laws but compensates for them by theorizing the existence of things such as "dark matter" and "dark energy," neither of which has ever been observed. For example, consider the lead paragraphs of a news report published in the *International Business Times*, March 7, 2015 titled "Does Dark Matter Originate From Higgs Boson? New Theory To Be Tested At CERN LHC." While the existence of dark matter is admitted to be only an inference which is based on a larger theory, by the second paragraph readers are told that dark matter exists and its origin only needs to be explained. Readers are told, without qualification, that something which has never been observed "constitutes 84% of the total matter in the universe."

> Dark matter has long remained one of the greatest unsolved mysteries of the universe. While its presence can be inferred from the gravitational pull it exerts on visible matter, the fact that it does not emit or absorb any radiation makes it next to impossible to detect.
>
> ...[I]n May this year, scientists will attempt to test a new model of particle physics -- one that attempts to explain the origin of the mysterious dark matter that constitutes over 84 percent of the total matter in the universe.

Picture yourself with a weight tied to a string that you are whirling about. If you let go of the string the weight will go flying away from you. The force pulling on the string is called centrifugal force. Stars in a galaxy of stars whirling about a center experience centrifugal force that ought to cause them to fly away from the center. But they don't. Something acts as "the string." Current theories must add vast amounts of a hypothetical mass, called dark matter, to explain why galaxies aren't torn apart by centrifugal forces. A Science Release (eso 1514) dated April 15, 2015 from the European Southern Observatory titled "First Signs of Self-interacting Dark Matter? Dark matter may not be completely dark after all" explained the problem:

> Our current understanding is that all galaxies exist inside clumps of dark matter. Without the constraining effect of dark matter's gravity, galaxies like the Milky Way would fling themselves apart as they rotate. In order to prevent this, 85 percent of the Universe's mass must exist as dark matter, and yet its true nature remains a mystery

## Dark Energy "Fact" Revised

As recently as 2011 three astronomers were awarded a Nobel Prize for determining that a theoretical force called dark energy was accelerating the [perceived] expansion of the universe. This went into all of the textbooks and became common wisdom in evolutionary cosmology. The paragraphs below, from a University of Arizona press release dated April 10, 2015, indicate the Big Bang story needs to be changed:

> The Nobel laureates discovered independently that many supernovae appeared fainter than predicted because they had moved farther away from Earth than they should have done if the universe expanded at the same rate. This indicated that the rate at which stars and galaxies move away from each other is increasing; in other words, something has been pushing the universe apart faster and faster.
>
> A UA-led team of astronomers found that the type of supernovae commonly used to measure distances in the universe fall into distinct populations not recognized before; the findings have implications for our understanding of how fast the universe has been expanding. Certain types of supernovae, or exploding stars, are more diverse than previously thought, a University of Arizona-led team of astronomers has discovered. The results, reported in two papers published in the *Astrophysical Journal*, have implications for big cosmological questions, such as how fast the universe has been expanding since the Big Bang. Most importantly, the findings hint at the possibility that the acceleration of the expansion of the universe might not be quite as fast as textbooks say. The discovery casts new light on the currently accepted view of the universe expanding at a faster and faster rate, pulled apart by a poorly understood force called dark energy.

## Why is Big Bang Cosmology Weird?

In the paragraph immediately above one reads that the currently accepted view is that the universe is being pulled apart by a force called dark energy. In the Science Release from the European Southern Observatory quoted earlier one read that "Without the constraining effect of dark matter's gravity, galaxies like the Milky Way would fling themselves apart as they rotate." To fit the Big Bang theory two undetectable opposing "dark" forces have been theorized: Dark energy is pulling the universe apart; dark matter keeps the galaxies together. If this seems weird, Jake Hebert, physics Ph. D., offered another perspective (Hebert, J. 2012. Why Is Modern Cosmology So Weird? *Acts & Facts*. 41 (8): 11-13):

> Cosmology is the study of the origin and structure of the universe. Because the Big Bang is the dominant cosmological model, most astronomers interpret all their observations to fit this paradigm.
>
> Big Bang cosmology is filled with a number of strange concepts, including *inflation, dark energy*, exotic forms of *dark matter,* and a *multiverse*. While valid scientific concepts such as quantum mechanics and relativity can indeed seem strange or counterintuitive, strange notions can also result from attempts to prop up a dying theory. Much of the weirdness of modern cosmology stems from an attempt to force the data to fit the Big Bang. Cosmology can be somewhat intimidating to non-specialists, but when one considers the reasons that Big Bang cosmologists invoke strange concepts like inflation, it quickly becomes apparent that the Big Bang is in trouble.

See also http://www.icr.org/article/8692/

John Hartnett, Ph.D., a physics and cosmology professor at the University of Adelaide in Australia explained why cosmology got weird:

This ludicrous situation has developed in astrophysics because of the initial assumption of *materialism* (matter and energy is all there is) and the dogmatic insistence that it must be rigorously applied to the origin and structure of this universe. As a result when physicists observe the rotation speeds of stars—not only in our own galaxy but also in many thousands of other spiral galaxies—they find that the stars in the spiral disks are moving too fast.

To fix this, the standard approach is to posit the existence, around every galaxy, of a spherical halo of dark matter that has just the right density, distribution and gravitational properties to solve the conundrum but neither emits nor interacts with electromagnetic radiation. Because astrophysicists cannot explain these high rotational velocities with standard tried-and-tested Newtonian physics, they have concocted the notion that galaxies really comprise between 80% to 90% dark matter—stuff that is everywhere but we cannot see or detect it by any method.

Beginning about 200 years ago, scientists started to abandon the Word of God as authoritative in such matters as the creation of the universe and hence it follows today that they believe in materialism—that there is no Creator and the universe just created itself from nothing. The alternative to accepting the materialists' explanation is to consider the possibility that the universe is not as old as they imagine (13.8 billion years) and that it was created only 6,000 years ago. For those fast stars this would mean they have not had time to fly apart.

For more on this subject see http://creation.com/has-the-dark-matter-mystery-been-solved

## Big Bang Bunk

It's amazing how many Catholic intellectuals have not merely tolerated the 'big bang' idea, but embraced it wholeheartedly. They brag that it was first proposed in 1927 by the Belgian Catholic priest Georges Lemaître. To hear their pronouncements, believers should welcome it as a major plank in our defense of the faith. "At last, we can use science to prove there's a creator of the universe." A good example of that thinking is Big Bang Jesuit Robert Spitzer with his video "Nothing to Cosmos: God and Science" and book *New Proofs for the Existence of God.* He's a "science entertainer" like Bill Nye, "the science guy" but in Roman collar instead of bow tie. The Big Bang hypothesis requires that most of the matter in the universe has to be something that has never been observed by natural science. Readers are urged to look up online "Inflation (cosmology)." Wikipedia has a long article from the evolutionists' perspective but from which one should certainly learn how absolutely theoretical and filled with inexplicable problems all of this is, even for those who believe it. The Big Bang proposes assumption-based hypotheses that are constantly subject to change to explain the pre-historic past by observations of the present. Despite what believers call them, these hypotheses are not scientific theories because they don't meet the criteria required to be a scientific theory. Scientific theories are testable and make falsifiable predictions. Sir Fred Hoyle readily saw through the fallacious assumptions behind the big bang theory. In 1994 he wrote, "Big-Bang cosmology refers to an epoch that cannot be reached by any form of astronomy, and, in more than two decades, it has not produced a single successful prediction." Explanations of the Big Bang must be taken on faith (called "science") while other faith-based explanations, such as "by the Will of the Uncaused Cause (God)," are disparaged as "religion." Faith in evolution-based theoretical cosmology thrives at taxpayer expense among grant-guzzling scientists in academia. The vast sums spent on theoretical cosmology by government are a misplaced priority. As an example of how theoretical cosmology is stoked for the bedazzlement of the general public by Big Bang story tellers in the popular media, consider "Ripples from the Big Bang," *NY Times*, March 24, 2014:

> When scientists jubilantly announced last week that a telescope at the South Pole had detected ripples in space from the very beginning of time, the reverberations went far beyond the potential validation of astronomers' most cherished model of the Big Bang.

Note how the writer slides from the "model of the Big Bang" in the paragraph above to the unqualified statement in the paragraph below that the "ripples" are in fact from the Big Bang and the reader is led to infer that the Big Bang has moved beyond speculation to being proven.

> The ripples detected by the telescope Biceps2 were faint spiral patterns from the polarization of microwave radiation left over from the Big Bang. They are relics from when energies were a trillion times greater than the Large Hadron Collider can produce.

And in the next paragraph, the *NY Times* writer does everything short of proclaiming that "the smoking gun has been found."

> These [radiation] waves are the long-sought markers for a theory called inflation, the force that put the bang in the Big Bang: an antigravitational swelling that began a trillionth of a trillionth of a trillionth of a second after the cosmic clock started ticking. Scientists have long incorporated inflation into their standard model of the cosmos, but as with the existence of the Higgs, proving it had long been just a pipe dream.

Within 6 months, this long-sought after sign of the presence of something supportive for the inflation theory (within the Big Bang theory) turned out to have been, literally, dust. An article entitled "Inflation, Elation, Deflation: Reflecting on BICEP2" on PBS, October 21, 2014, recounted how

> Six months ago astrophysicists working on an experiment called BICEP2 were celebrating what some called the discovery of the century: the detection of a specific polarization signature in the cosmic microwave background radiation that, interpreted conservatively, provided the most direct confirmation ever of cosmic inflation. Read more expansively, it was seen as evidence for the quantization of gravity and the existence of the multiverse.

The PBS article continued to report that:

> Last month, new data released by the Planck team confirmed that all or most of the BICEP2 signal could indeed be due to dust. It doesn't rule out the possibility that BICEP2 saw something real, but shows that the signal can't yet be untangled from the noise.
>
> All of which has scientists and science media wringing their hands over what—if anything—they should have done differently. The splashy announcement, accompanied by literal and figurative champagne-cork-popping, as we covered here, coincided not with publication in a peer-reviewed journal but with the publication of results online. Should the authors have waited for peer review to announce their results? Should journalists have been more circumspect?

Indeed, and should Catholics also be more circumspect before reinterpreting *Genesis* to fit the claims of the evolutionists?

## Truth Doesn't Matter to True Believers

In *The Doctrines of Genesis 1-11*, scientist-priest Victor Warkulwiz notes that "Theorists can offer no physical reason for the big bang. This is a serious weakness of the theory." He quoted Big Bang theory proponent Steven Weinberg, author of *The First Three Minutes of the Universe: A Modern View of the Origin of the Universe* as writing that "There is embarrassing vagueness about the beginning." In his book Fr. Warkulwiz identifies and explains 16 deficiencies of the Big Bang theory.

## Anything Qualifies as Science

There is almost no end to the number of bizarre statements emanating from taxpayer-funded grant and contract chasers that become part of our popular culture. They even have a theory that says the universe should not exist. This was reported in June 23, 2014, by Livescience.com, an evolutionary cosmology and biology promoting website. The article was "Universe Shouldn't Be Here, According to Higgs Physics" It reported that:

> The universe shouldn't exist — at least according to a new theory. Modeling of conditions soon after the Big Bang suggests the universe should have collapsed just microseconds after its explosive birth, the new study suggests.

The evolution-promoting media writers treat theories as if they were facts, just as school textbook writers do. For example, the Big Bang is only a current working hypothesis of theoretical physics that seeks a naturalistic explanation for the universe's existence. But in the article cited above, the writer treats the "explosive birth" as a fact and all speculations qualify as science as long as they don't include God. A great article regarding the big bang is "Secular Scientists Blast the Big Bang." Read it at http://creation.com/secular-scientists-blast-the-big-bang.

## 18th Century Science Still Being Taught

In the excerpt below, from a report on Space.com, February 20, 2015, note how "clouds of gas" explain the origin of the sun and stars.

> Earth's water has a mysterious past stretching back to the primordial clouds of gas that birthed the Sun and other stars. By using telescopes and computer simulations to study such star nurseries, researchers can better understand the cosmic chemistry that has influenced the distribution of water in star systems across the Universe.

"Primordial clouds of gas birthed the sun and other stars." That is the "nebular gas hypothesis" formulated by 18th Century German philosopher Immanuel Kant in his *Universal Natural History and Theory of the Heaven.* Three centuries later that is taught in schools as a fact simply because nothing better has come along and Kant's hypothesis doesn't involve God. Stars, galaxies and our solar system supposedly formed from gas and dust clouds clumping together, rather than dispersing. But no model can successfully simulate this.

Read https://creation.com/the-naturalistic-formation-of-planets-exceedingly-difficult. Also read http://creation.com/stars-dont-form-naturally And http://www.icr.org/article/10347/

For almost every solar system body the magnetic field strength is a surprise. Mercury shouldn't have a magnetic field (but it does); surely Venus and Mars should have one like ours (but they don't); Jupiter's shouldn't be so strong; Saturn's shouldn't be so symmetrical; and Uranus' and Neptune's shouldn't be so *a* symmetrical. The geological behavior is frequently unexpected, too (volcanism on bodies too small to retain their heat for billions of years—Io, Pluto, Charon, and more). Essentially, the preferred naturalistic models for the development of our solar system cannot account for any of its major features.

For a debunking of the evolutionists' explanation of our solar system see "Our Solar System: Evidence of Creation" youtube.com/watch?v=s9_o7NGTkJc.

**Black Holes**

"Black holes" are areas of mass in space that have such a gravitational pull that it is said that not even light could escape from them. Stephen Hawking is a Cambridge University theoretical physicist and cosmologist who helped to popularize black holes in the 1970s. Hawking is an icon in the evolutionary world. Popular media outlets hang on his every word and have made him a celebrity. He regularly conjures up new theories and it doesn't seem to matter to his fans how speculative his theories are. In early 2014, evolutionary cosmology was set all a-twitter when Hawking published a paper on the internet that upset the scientific consensus. It was no surprise then that *Yahoo! News*, in response to Hawking's paper, carried this headlined story, "Stephen Hawking says black holes don't exist."

Actually, extremely dense areas which have been named "black holes" do exist in space beyond our solar system. An important real science finding was reported in the February 11, 2016 issue of *Physical Review Letters.* These "holes" move around. Einstein's general relativity theory predicted that if two of them merged it would create a transient gravitational wave signal. Years ago, at a cost of $1.1 billion to U.S. taxpayers, elaborate equipment was set up in Louisiana and Washington State to test that prediction when and if a merger took place. On September 14, 2015 a transient gravitational wave signal was detected. Based on equations that relate gravitational force to mass, the mass of one "hole" was estimated at 36 times that of our sun and the other at 29 times that of the sun. After merger the mass was estimated at 62 times that of our sun and a mass estimated at 3 times that of our sun was radiated away as gravitational waves.

## The Uncaused Cause

Many Catholics are familiar with one of the logical arguments for the existence of God, namely, cause and effect. An article by Bruce Gordon called "What Should We Make of Gravity-Wave Detection?" published on evolutionnews.org explained how that gravity wave experimental result points to an intelligent Creator.

> What this discovery really provides is additional and exceedingly strong confirmation of Einstein's already well-confirmed theory of general relativity by directly establishing the existence of gravity waves and giving further evidence of the existence of black holes. The significance of discoveries confirming general relativity relate to one of the implications of the theory itself. As Roger Penrose and Stephen Hawking demonstrated in the late 1960s, regardless of which solution of Einstein's equations is embraced, all backward-traced spacetime geodesics in classical general relativity terminate in a singularity, implying that space-time, matter, and energy all came into existence at some point in the finite past. This, of course, is the essence of [materialist] Big Bang cosmology.
>
> In other words, the universe began to exist, and there is no physical explanation in cosmology or physics for why this happened. This opens the door to various cosmological arguments, including, of course, the Kalam [cosmological] argument: everything that begins to exist has a cause and the universe began to exist, therefore the universe has a cause. This cause is obviously not something physical, because it transcends the physical universe, and it is obviously not something abstract and mathematical, because even if one is a Platonist and thinks such entities have mind-independent existence, they are causally inert and impotent. The only causal option left is an immaterial transcendent personal agent of

immense power and wisdom. So the effect of this new discovery is really just to strengthen theistic arguments that draw on general relativity itself.

### Creation Cosmology

Henry Morris has noted that one of the traditional "discrepancies" attributed by the skeptics to the *Genesis* account of creation is the fact that there was "light" (Hebrew *or*) on the first day of the creation week, whereas God did not create the "lights" (Hebrew *ma-or*) to rule the day and the night until the fourth day. Morris pointed out that evolutionary cosmologists find no problem in having light before the sun. According to their speculative reconstruction of cosmic history, light energy was produced in the imaginary "Big Bang" 15 billion years ago, whereas the sun "evolved" only five billion years ago. The fact that light is an entity independent of the sun and other heavenly bodies is one of the remarkable scientific insights of the Bible. As the basic form of energy it is significant that the first recorded word spoken by the Creator was: "Let there be light" (*Genesis 1:3*). Some Christians have been led to accepting a 15 billion-year old cosmos because of the so-called "starlight problem." How could light from stars recently created be visible when they are millions of light years distant from Earth? A similar question plagues the Big Bang Theory: "How is it possible that areas of the universe 20 billion light years apart are at the same temperature?" That is known as "the horizon problem" and has caused many evolutionary cosmologists to abandon the big bang model. Secular astronomers have proposed many possible solutions to it, but no satisfactory one has emerged to date. The horizon problem is explained here http://creation.com/light-travel-time-a-problem-for-the-big-bang

## As Much Information As You Want

This book is not intended as a book of science. It is to stimulate further study. Readers who wish more scientific information on these issues should read "Life of a Universe: Part 1 Creation http://creation.com/life-of-a-universe-creation, Solar System and Extra-solar Planets http://creation.com/solar-exoplanet-qa, and Age of the Earth http://creation.com/age-of-the-earth and http://creation.com/cherry-lewis-the-dating-game-one-mans-search-for-the-age-of-the-earth-book-review

## Believe Divine Revelation Instead of Fallible Men

Earlier in this chapter, this writer referred to the reverence accorded to Stephen Hawking, no matter what he says. In a book Hawking co-authored, *The Grand Design,* published in 2010, one finds his opinion that

> It is not necessary to invoke God to light the blue touch paper and set the universe going. Instead, the laws of science alone can explain why the universe began. Our modern understanding of time suggests that it is just another dimension, like space. Thus it doesn't have a beginning. Because there is a law such as gravity, the universe can and will create itself from nothing. Spontaneous creation is the reason there is something rather than nothing, why the universe exists, why we exist.

Logic doesn't seem to be his strong point; 'self-creation' is self-contradictory. Something can do something—including create—only if it exists; something not yet existing has no power to do anything, *including create itself.* In asserting that time had no beginning and the universe created itself from nothing, Hawking is preaching the first dogma of Humanism. One does not have to be a famous scientist to tell us that the universe created itself from nothing. Non-scientist Humanist philosophers, such as John

Dewey, told the world the same thing in 1933 when they published *Humanist Manifesto I.* The Humanists described themselves as a new religion, that is, a religious movement meant to transcend and replace deity-based religions:

> While this age does owe a vast debt to the traditional religions, it is none the less obvious that any religion that can hope to be a synthesizing and dynamic force for today must be shaped for the needs of this age. To establish such a religion is a major necessity of the present. It is a responsibility which rests upon this generation. We therefore affirm the following:
>
> First: Religious humanists regard the universe as self-existing and not created.

Modern scientific instruments and space exploration have disproved a crucial evolutionary principle, namely, Uniformitarianism. Uniformitarianism is the assumption that the same natural laws and processes that operate in the universe now have always operated in the universe in the past and apply everywhere in the universe. The YouTube video, called "Our Solar System: Evidence of Creation" referred to earlier gives many examples of how space discoveries have gone against predictions based on uniformitarian principles. (The inseparable dependence of biological evolution theories on Uniformitarianism is explained in chapter six.)

**The New Gospel**

Fr. Warkulwiz observed in *The Doctrines of Genesis 1-11* that "The doctrine of an ancient cosmos is asserted and proclaimed as a fact so often in scientific presentations, even when the context doesn't call for it, that it becomes obvious that a "gospel" is being preached." I have just identified for the reader the name of that "gospel," namely, *Humanist Manifesto I* which has been the practical Creed of the American education industry even if many or most educators never heard of *Humanist Manifesto I* (1933) or *Humanist Manifesto II* (1973).

Fr. Warkulwiz went on to explain that "It is the gospel of naturalism, in which true religion has no say; and the continual assertions are professions of faith in that gospel. The antibiblical doctrine of an ancient cosmos is a fundamental tenet of that faith, and it is a doctrine held to be true beyond question."

When Stephen Hawking gives his opinion that the universe created itself from nothing, it is automatically assumed that his uttering is science; therefore, his Humanist doctrines may be taught in public schools. If a creation-supporting scientist said the universe appears to have been designed and that implies a Designer that would be considered "just religion" and a "violation of Church and State separation" a "doctrine" invented by Freemasons on the Supreme Court.

**God Does Not Deceive**

The big question, then, is: "Why do Christians put up with this takeover of our taxpayer-funded institutions?" Could the answer be that too many Christian clergy and lay intellectuals have accepted evolution, the philosophical basis of Humanism, and won't let go of it? Because so many have, the ordinary Catholics have no teachers for reasons explained later in this book by Cardinal Joseph Ratzinger. Fr. Warkulwiz compares those who reject *Genesis* in favor of scientism to Eve.

> God does not deceive us. We deceive ourselves if we chose to ignore His Word and seek the truth elsewhere. Those who reject the Genesis account and seek information about the origin of the world from natural causes alone follow the example of Eve who would not believe what God told her but instead sought truth from the tree of the knowledge of good and evil.

# Chapter 4-Evolutionary Biology

Moving on from the evolution-based Big Bang cosmology to the biological evolution of humans, *The Wanderer* writer Fitzpatrick explained what he learned from his Catholic religious order teachers as a youth.

> I don't know if my experience is typical, but this is the understanding of the Book of *Genesis* that I have been taught since I was in high school in the Bronx in the 1950s. The Marist Brothers who taught me at that time would tell their students that Catholics are free to believe that evolution took place, as long as they understood it to be a process begun by God, and one in which human beings were created when God infused a soul into the evolving creature that became man. This was the same understanding taught to me by Jesuit priests at Fordham in the 1960s.

It is not surprising that the above non-scientific, non-Biblical explanation would have been taught by Jesuits given the present state of the Jesuits in which a good number of them push beyond the limits of what is acceptable Catholicism. The spiritual death spiral of the Jesuits that began 100 years ago was chronicled in the 1980s by James Hitchcock in *The Pope and the Jesuits: John Paul II and the New Order in the Catholic Church* and in Malachi Martin's *The Jesuits: The Society of Jesus and the Betrayal of the Roman Catholic Church.*

Nor is it unusual that their students accepted it when that is what they were taught. Some Jesuits "got aboard," so to speak, with evolution long before evolution's most famous early expositor, Charles Darwin, was born. For example, in the late 18th Century, English Jesuit John Needham was the leading voice arguing that life could spontaneously arise from non-life. His views were based on his private interpretation of a verse in *Genesis* 1. A French scientist, Louis Pasteur, disproved life from non-life by a series of experiments in 1861; the origin of life could not be explained as theorized by Darwin in 1859.

**Life from Non-Life Speculation Never Dies**

Life arising from non-life is part of the faith package that comes with evolution. Fr. Needham would have rejoiced to hear Darwin suggesting in 1871 that the original spark of life may have begun in a "warm little pond, with all sorts of ammonia and phosphoric salts, light, heat, electricity, etc. present, so that a protein compound was chemically formed ready to undergo still more complex changes." That speculation is the origin of the "primordial soup" explanation for the beginning of life found in so many school textbooks and nature programs on PBS. The Primordial Soup Hypothesis was resurrected in 1936 by Russian chemist A. I. Oparin. He proposed how it could have happened if conditions on the earth back then (whenever back then was) were different than they were at present. Among other things, the proposed soup had to be in an oxygen-free atmosphere. The beauty of that speculation from the evolutionist view point is that it can be told to children without any need to prove it. And it can't be disproved.

In 1952, a graduate student, Stanley Miller, and his professor tested Oparin's idea by mixing water and three gases in an oxygen-free environment, ran electricity through the mix and produced two amino acids. These are not alive but are chemical compounds integral to protein. That was the famous Miller-Urey experiment. So constantly repeated is the propaganda regarding the importance of that lab experiment that this writer has a friend with two science degrees who told him that that experiment had proved life can come from non-life. In 2000, Miller was working for NASA and trying to find ways to rescue the original scheme. For more on Miller-Urey see http://creation.com/life-in-a-test-tube . Also https://creation.com/origin-of-life-research

**The PBS Evolution Project**

A classic example of the "Life from Non-Life" propaganda is the PBS Evolution Project. The Project includes a seven-part television series, a web site, a multimedia library, and an educational outreach program. The TV series is "Evolution" and it was produced and first broadcast circa 2000. The companion book to the PBS Series is *evolution: The Triumph of An Idea.* (Why they chose to not capitalize "evolution" in the title was not stated.) It is interesting that it was called the "triumph of an idea" rather than a "triumph of science." It supports this writer's contention that evolution is more about faith than science.

The importance evolutionists attached to the PBS Evolution Project is manifest by who was chosen to write the TV program's companion book's six-page Introduction. It was Stephen Jay Gould, the most famous evolutionist in America. He was Professor of Zoology and Professor of Geology at Harvard and the curator for invertebrate paleontology in that university's Museum of Comparative Zoology. He was at Harvard from 1967 until his death in 2002. As of 2002 he had published 22 books. He was also America's greatest communicator of evolutionary ideas to the ordinary laymen which he accomplished through more than 300 essays in *Natural History* magazine between 1974 and 2001. He was the consummate story spinner for he wrote interesting and captivating prose. In his Introduction to *evolution: The Triumph Of An Idea,* Gould started with an apocryphal story making fun of the wife of a Church of England clergyman. According to this story which took place in the "early days of Darwinism" the woman exclaimed to her husband when she grasped the scary novelty of evolution: 'Oh my dear, let us hope that what Mr. Darwin said is not true. But if it is true, let us hope that it will not become generally known!'

Then, despite the universities, public schools, media, and Federal Government dedicated to evolution propaganda Gould whined that evolution had not become generally known in the United States:

> For what Mr. Darwin said is clearly true, and it has also not become generally known (or, at least in the United States, albeit uniquely in the Western world, even generally acknowledged). We need to understand the reasons for this exceedingly curious situation.

As noted earlier in this book, Pew Research indicates that as of 2014, 65% of adult Americans believe evolution is a fact. To see why not everyone is fooled, read "Can Darwinian Evolutionary Theory Be Taken Seriously?"
http://natureinstitute.org/txt/st/org/comm/ar/2016/teleology_30.htm
*Evolution: The Triumph of An Idea* begins with a narrative of Darwin's history, the compatibility of his thought with that of his many contemporaries, and the ridicule by PBS of those who disagreed with him. And those who disagreed were more numerous. For example, Scotsman James Clerk Maxwell (1831-1879) was one of the greatest scientists who have ever lived. In presenting a paper, """Discourse on Molecules." to the British Association at Bradford in 1873 he pointed out that

> No theory of evolution can be formed to account for the similarity of molecules, for evolution necessarily implies continuous change .... The exact equality of each molecule to all others of the same kind gives it ... the

essential character of a manufactured article, and precludes the idea of its being eternal and self-existent.

PBS is correct that Darwin's ideas became the centerpiece of a naturalistic scientific consensus that has culturally [if not scientifically] triumphed. The PBS evolution extravaganza provides an opportunity to illustrate life from non-life propaganda by reference to where it was transferred into print in the companion book. That is in the book's section, "In Search of Life's Origins" (pages 104-115).

### Life from Space Debris

On the first page is an artist's conception drawing of a bright object moving against a background of what looks like a night sky full of stars. The caption on the picture is "Comets may have seeded the early Earth with many of the building blocks of life." That speculation would be called a scientific theory by evolutionists. When interviewed by Ben Stein in the movie *Expelled,* Richard Dawkins said that space aliens from a technically advanced civilization may have put the building blocks on the comet. The term "building blocks" is a scientific trivialization typical of evolutionary story telling. There are no "building blocks" to be stacked up or mixed up to produce life. Life is on a completely different order from the components that make it up. A child watching this on TV might accept seed-bearing comets as science. Should we?

### Life from the Hot Tub

On the next page there is a truly beautiful picture of a hot spring at Yellowstone Park surrounded by snow just at dawn. The caption is "Hot springs in Yellowstone are home to some of the most primitive microbes on Earth. Researchers suspect that life may have begun 4 billion years ago in near-boiling water." That's Darwin's "warm little pond" speculation from 1871 updated with 21$^{st}$ Century audio-visual effects. Who would those unnamed researchers be and what observed data supports their suspicion? Again one can see evolutionary story telling in the term "primitive microbes." Two meanings for the word "primitive" are: (1) of, belonging to, or seeming to come from an early time in the very ancient past and (2) very simple and basic, made or done in a way that is not modern and that doesn't show much skill. One wonders, if they are from the very ancient past why haven't they evolved into fish by now? Or, if they are very simple and basic, how did they become alive and then get all of the genetic material to become our ancestors? All evolutionary writing is aimed at communicating more than is actually written. In this caption, it doesn't actually say that the microbes in the spring are evidence of anything at all. But with the beautiful picture and the professional narrator, what child would not infer what the script writer wanted to be inferred?

**The Space Debris Assembles Itself**

The text on the pages is equally pseudo-science. For example,

> The first step in the rise of life was to gather its raw materials together. Many of them could have come from space. Astronomers have discovered a number of basic ingredients for life on meteorites, comets, and interplanetary dust.

The above asserts that life arose in a series of steps. Unstated is who or what is the active agent taking the "first step" of gathering raw materials together because there was no life. The evolution story, like all false ideologies, begins with a bald assertion which is then treated as a fact upon which the story is built. From here on there will be a lot of "could have" and "may have" speculation statements that build on the initial bald assertion.

> As these objects fell to the early Earth, they could have seeded the planet with components for crucial parts of the cell, such as the phosphate backbone of DNA, its

> information-bearing bases, and amino acids for making proteins.

One wonders how the information got into the "information-bearing bases." Only intelligence can produce such information.

> As these compounds reacted with one another, they may have produced more life-like forms.

That sentence puts it out there for belief of the gullible that non-living things falling to earth could have produced life-like forms. The term "life-like forms" is jargon that means nothing.

> Chemical reactions work best when the molecules involved are crowded together so they bump into one another more often; on the early Earth, the precursors of biological matter might have concentrated in raindrops or the spray of ocean waves.

The assertion above is leading to an assertion to be made later in this fantasy story that non-biological material evolved into biological material by chemical reactions.

> Some scientists suspect that life began at the midocean ridges where hot magna emerges from the mantle. The branches nearest to the base of the tree of life, they point out, belong to bacteria and archaea that live in extreme conditions such as boiling water or acids. They may be relics of the earliest ecosystems on the planet.

"Some scientists suspect" is one of the most hackneyed phrases of the evolution genre. It conveys the aura of scientific knowledge and authority but doesn't actually have anything behind it. The paragraph above doesn't actually say how the bacteria and archaea came to life but a child could infer more than has actually been said. A college student might ask: "Do you suspect, professor, that if something is not living and you boil it in hot magna it will become a living thing?" If he says "yes" the student ought to ask for a tuition refund.

> Scientists suspect that prebiological molecules became organized into cycles of chemical reactions that could sustain themselves independently. A group of molecules would fashion more copies of itself by grabbing other molecules around it.

Are the vague "prebiological molecules" cousins of the vague "precursor biological material" mentioned above? How can molecules organize? One of the fundamental laws of science, the Second Law of Thermodynamics, is that material tends toward disorganization. How could lifeless molecules know how to grab molecules around them, or which ones to grab, to make copies of themselves? How would they "know" when the grabbed molecules constitute a copy of themselves?

The next paragraph is asserting chemical to biological evolution.

> There may have been many separate chemical cycles at work on the early Earth. If they used the same building blocks to complete their cycles, they would have competed with each other. The most efficient cycle would have outstripped the less efficient ones. Before biological evolution, in other words, there was chemical evolution.

That last sentence doesn't actually say that biological life resulted from chemical evolution but that is what was implied because that is part of the evolutionist story telling for the public. Almost no mainstream evolutionary biologist proposes chemical evolution as the source of first life. In the next chapter of this book you will read that an evolution-believing chemist of note labels that as "astonishingly improbable."

> Ultimately, these molecules gave rise to DNA, RNA, and proteins.

Ultimately, the tooth fairy came, took my tooth and left DNA under my pillow. It's as simple as that. "Ultimately" according to evolutionists, the most complex things just "arise."

### Amazing Information-Rich DNA

DNA is so amazingly complex that this writer will only touch on its importance. As much as anyone would ever want to know can be found in *Darwin's Doubt* by Stephen Meyer. Meyer reveals from the research of modern science a cellular complexity that was undreamed of 50 years ago when the evolution myth with its simplistic story became educational orthodoxy in America. The PBS tooth fairy science story said "information-bearing bases" could have been included in the "basic ingredients for life on meteorites, comets, and interplanetary dust." Those "information-bearing bases" of DNA are the genetic code. Information is created only by an intelligent mind. Complexity alone does not necessitate intelligence, but something called *specified complexity* does. Whereas simple complexity denotes mere intricacy, specified complexity occurs in a configuration when it can be described by a pattern that displays a large amount of independently specified information and is also complex, which is defined as having a low probability of occurrence. In other words, there is of necessity a very specific pattern or complexity that must occur in the configuration. Genetic code provides a classic example of this. Individual genes must be arranged in very specified patterns. Every gene needs to be located in a specific place for them to function properly. The configuration is not just complex but also specified. If the re-discovery and validation at the beginning of the 20th Century of genetic laws demonstrated by Mendel in the 1860s gave Darwinism a kick in the teeth, the discovery of DNA in 1953 would have buried it if any of this had anything to do with science. Darwinism, in one or more of its latest syntheses, can no more be removed from Humanist religion than the Incarnation can be removed from Christianity. An article on Nobelprize.org called, "The Discovery of the Molecular Structure of DNA-The Double Helix," explained the importance of that discovery as follows:

> The sentence "This structure has novel features which are of considerable biological interest" may be one of science's most famous understatements. It appeared in April 1953 in the scientific paper where James Watson and Francis Crick presented the structure of the DNA-helix, the molecule that carries genetic information from one generation to the other.

In the decades that followed, much genetic research has been completed. The most important of that is sponsored and funded by the National Institutes of Health's National Human Genome Institute. On the page devoted to Deoxyribonucleic Acid (DNA) there are questions and answers that one can read:

> What is DNA?
>
> We all know that elephants only give birth to little elephants, giraffes to giraffes, dogs to dogs and so on for every type of living creature. But why is this so?
>
> The answer lies in a molecule called deoxyribonucleic acid (DNA), which contains the biological instructions that make each species unique. DNA, along with the instructions it contains, is passed from adult organisms to their offspring during reproduction.

Does that sound like a description of the mythical evolutionary "Tree of Life" that is in school textbooks illustrating the evolution from bacteria to you? In 2005 science even discovered that there is an epigenetic code behind the DNA code! The more science learns about epigenetic and DNA information the more troublesome it becomes for evolutionists to tell their fables. A great article called "The Genetic Puppeteer" about how the epigenetic code controls the DNA code is at http://creation.com/the-genetic-puppeteer.

**Primordial Soup on the Universities' Menu**

Prestigious universities are still ladling out the primordial soup story to gullible students along with the "Tree of Life" fantasy. For example, the U. of California at Berkley had online in 2016 "Understanding Evolution: your one-stop source for information about evolution." Evolution 101 features a learning module called "From Soup to Cells- The Origin of Life." It is similar to the tooth-fairy science spun by PBS in 2000.

**Slow, Medium or Fast DNA**

Articles in science journals indicate the problems that evolutionary geneticists are having with the so-called "DNA clock." An article at arstechnica.com published March 28, 2013, "Fossil DNA used to reset humanity's clock," was a report of research published in *Current Biology*. According to this article:

> Some time in humanity's past, a small group of Homo sapiens migrated out of Africa before spreading out to every possible corner of the Earth. All the women of that group carried DNA inherited from just one woman, commonly known as mitochondrial Eve, whose DNA was inherited by all humans alive today. But the exact timing of this migration is not clear, and it has sparked debate among geneticists. Now, new research published in *Current Biology* may help calm both sides.

*[See "You and the Wives of Noah's Sons" in chapter 6 to learn how the DNA of "mitochondrial Eve" branched out.]*

The evolution-based story was accompanied by a picture of a painting showing three tall, naked black people, with ape-like heads walking, and the third one's legs had the bowed-legged look of a gorilla. That picture reflects Darwin's racial theories. The full title of his famous 1859 book is *The Origin of Species By Means of Natural Selection and the Preservation of the Favored Races in the Struggle for Life*. Sanger, Hitler, Marx and others based their racial theories and coercive eugenics on Darwinism. The medical literature of the early 20th Century based on Darwinism contributed so much to the racism in America in that period. (For more on that see http://creation.com/the-darwin-effect-review)

The *Current Biology* article explained that evolutionists use "molecular clocks" based on "changes in DNA that accumulate over time" and to accurately calibrate the "clock" it is necessary to accurately measure the rate of mutations. In 2012, UK researchers published a new, lower rate of mutations.

> Based on their results, it would seem that human DNA may change much more slowly than was previously thought. The slow mutation rate puts the date of human migration out of Africa at somewhere between 90,000 and 130,000 years ago.

But that caused a problem based on what other scientists think they know. A German researcher was quoted as saying that, "This contradicts what we know from fossil studies." The German researcher, whose results were published in *Current Biology*, reported a new mutation rate that was higher than the UK researchers' rate but lower than the older rate that was commonly accepted. According to the Germans, that fantasized trek out of Africa "has been set to 62,000 to 95,000 years ago." The German added that it would be no surprise "if the precise date of migration was later than archaeologists currently believe, but it certainly is earlier than what the UK researchers claim."

Fast forward to a February 2015 conference described in a March 10, 2015 article in *Nature*, international weekly journal of science, "DNA mutation clock proves tough to set: Geneticists meet to work out why the rate of change in the genome is so hard to pin down." Without actually saying so, the article illustrated how evolutionists are completely flummoxed by DNA research. According to the article, geneticists are having trouble deciding between one measure of how fast human DNA mutates and another, which is half that rate.

> The rate is key to calibrating the "molecular clock" that puts DNA-based dates on events in evolutionary history. So at an intimate [Human Mutation Rate] meeting in Leipzig, Germany, on 25–27 February, a dozen speakers puzzled over why calculations of the rate at which sequence changes pop up in human DNA have been so much lower in recent years than previously. They also pondered why the rate seems to fluctuate over time.

**Pre-DNA Assumptions**

It seems that as far back as the 1930s evolutionists put a number on the so-called "human mutation rate," and that was decades before DNA was identified as one of the carriers of encoded genetic information. This was the state of science that informed the teachers of Mr. Fitzpatrick and the teachers of generations after them, as they parroted evolutionary time-scales as facts. The article reported that the new research in the last six years showed a "human mutation rate" about half of the rate that had been the scientific consensus. This really messes up the evolutionary "tree of life," descent from common ancestor fiction. An article, also in *Nature* and published in September 2012, explained why:

> Although a slowed molecular clock may harmonize the story of human evolution, it does strange things when applied further back in time," says David Reich, an evolutionary geneticist at Harvard Medical School in Boston, Massachusetts. "You can't have it both ways. For instance, the slowest proposed mutation rate puts the

> common ancestor of humans and orangutans at 40 million years ago," he says: more than 20 million years before dates derived from abundant fossil evidence. This very slow clock has the common ancestor of monkeys and humans co-existing with the last dinosaurs. "It gets very complicated," deadpans Reich.

Because of this, the March 2015 article reported that because they are "reluctant to abandon the older numbers completely, many researchers have started hedging their bets in papers, presenting multiple dates for evolutionary events depending on whether mutation is assumed to be fast, slow or somewhere in between." Is it science when scientists start substituting variables for concrete numbers depending upon the result they want?

**Bad and Uncertain**

One of the organizers of the February 2015 conference, David Reich, the population geneticist from Harvard, presented the results of two recent studies which had calculated a "slow rate" and another which had calculated an "intermediate rate." He said he was unable to explain the difference and at that point he injected the only moment of reality into the conference when he was quoted as having said

> The fact that the clock is so uncertain is very problematic for us...It means that the dates we get out of genetics are really quite embarrassingly bad and uncertain.

Evolutionists "know" humans have been around for hundreds of thousands of years. Therefore "much of the meeting revolved around when it accelerated and decelerated — and why." Have they ever considered that maybe humans have not descended from anything and have only been here for 6000 years or so?"

## Mutations Are Winding Us Down

The most puzzling part of the genetic "human mutations rate" discussion is the mixture of the theory that humans descended from some non-human ancestor (meaning that they added genetic information to become a different species through mutations), with the real observations that humans only develop mutations with harmful effects and pass them on to following generations (meaning that humans are becoming more genetically flawed). Calling someone a mutant is an insult because mutations, which are copying mistakes in DNA, are almost always bad. In fact, many mutations are known by the diseases they cause. In *Genetic Entropy and the Mystery of the Genome* former Cornell University professor Dr John Sanford pointed out the seriousness of this problem. He showed that mutations are rapidly decaying the information within the human genome. According to evolutionary theory, mutations coupled with natural selection, is the means by which new information arises. But, according to Sanford, if mutation and selection cannot preserve the information already in the genome it's difficult to imagine how it created all that information in the first place.

Beside the evolutionary geneticists, that February 2015 meeting discussed above drew researchers with an interest in cancer and reproductive biology — fields in which harmful mutations have a central role. In fact, it is the bad mutations, that is, the genetic flaws being passed along from generation to generation that are of scientific interest. The May 2015 issue of *Discover*'s cover story was "Evolution Gone Wrong: Why Humans Struggle to Adapt to Modern Diseases." The article is full of evolutionary bravado, including assuring readers that "humans are still evolving," but featuring the concerns of Harvard evolutionary biologist Daniel Lieberman who has coined the new term, dysevolution, to describe the degenerating state of human health as disease is passed through inheritance. For example, a mutation in the FBN1 gene is associated with life-threatening aortic aneurisms and children have a 50% chance of inheriting it from an affected parent. For information regarding what mutations are and what they do there is a short tutorial online at icr.org, "Mutations: The Raw Material for Evolution?"

### A Constantly-Shifting Contradictory Fairy Tale

*Scientific American*'s September 2014 Special Evolution Issue provided an article "The Human Saga: Evolution Rewritten" that said "awash in fresh insights, scientists have had to revise virtually every chapter of human history." The scientists to whom the article refers are not empirical scientists; they are "pre-historic" scientists who dig up fossils and make interpretations according to *a priori* strictures. Those strictures require that human descent from animals is a fact. The article features the usual artist drawing of the supposed "Human Family Tree" of skull fragments, all distinct with no transitional forms, and this commentary on that imagined "tree."

> With relatively few fossils to work from, scientists' best guess was that they all could be assigned to just two lineages, one of which went extinct and the other of which ultimately gave rise to us. Discoveries made over the past few decades have revealed a far more luxuriant tree, however-one abounding with branches and twigs that eventually petered out. This newfound diversity paints a much more interesting picture of our origins but makes sorting out our ancestors from the evolutionary dead ends all the more challenging, as paleoanthropologist Bernard Wood explains in the pages that follow.

On the following pages the story notes that because of all of the fossils found in the last 40 years "figuring out how they are all related—and which one led directly to us—will keep paleontologists busy for decades to come." In other words, they haven't a clue but as long as they have a theory, it is "science."

## Warning for Would-Be Polygamists: It Is Hard Work

The stuff that appears in tooth-fairy science magazines can be fun. For example, the above mentioned Special Evolution Issue of *Scientific American* also featured an article called "Powers of Two." The article describes a study by scientists who say they discovered how monogamy evolved in human culture. According to anthropologist C. Owen Lovejoy of Kent State University

> Soon after the split from the last common ancestor between the great ape and human evolutionary branches more than seven million years ago, our predecessors adopted a transformative trio of behaviors: carrying food in arms freed by bi-pedal posture, forming pair bonds and concealing external signals of female ovulation. Evolving together, these innovations gave hominins, the tribe that emerged when early humans diverged from chimpanzees a reproductive edge over apes…an ancestral polygamous mating system was replaced by pair bonding when lower-ranked hominin males diverted energy from fighting toward finding food to bring females as an incentive to mate. Females preferred reliable providers to aggressive competitors and bonded with the better foragers. Eventually the females lost the skin swelling or other signs of sexual receptivity that would have attracted different males while their partners were off gathering food.

This "science" covers two pages in the text but was summed up by three sentences in very large type:

> Keeping many mates is hard work. It involves a lot of fighting with other males and guarding females. Monogamy might have emerged as a way to reduce the effort.

Perhaps the Kent State anthropologist discovered how evolution happened by "survival of the laziest."

# Chapter 5-Why That Was Important

The reader may have been bored reading the PBS fantasy "In Search of Life's Origins," and other evolutionary tales but this writer felt it was necessary. One purpose of this book is to encourage Catholics to have more confidence in God's Revelation and the guidance of the Holy Spirit given to Popes exercising their Ordinary Magisterium. The Humanists abuse us and run the country based on having convinced many of us that science and history are on their side. Make no mistake, what you just read is the best explanation Humanists have for the origin of life and human history and everything that modern science is learning challenges that nonsense.

**How Did Life Begin? "We Need a Really Good New Idea"**

In the last chapter I showed how the PBS Evolution Project suggested that life evolved from chemicals. To appreciate how evolution propaganda is so far removed from science consider what a preeminent chemist who doesn't believe God created life said about that. The American Chemical Society (ACS) each year awards a gold medallion called the Priestly Medal to recognize distinguished services to chemistry. In 2007 the winner was George M. Whitesides and the ACS's journal, *Chemical & Engineering News* reported the address he gave the ACS on that occasion. Regarding the origin of life he said:

> This problem is one of the big ones in science. It begins to place life, and us, in the universe. Most chemists believe, as do I, that life emerged spontaneously from mixtures of molecules in the prebiotic Earth. How? I have no idea. Perhaps it was by the spontaneous emergence of "simple" autocatalytic cycles and then by their combination. On the basis of all the chemistry that I know, it seems to me astonishingly improbable. The idea of an RNA world is a

good hint, but it is so far removed in its complexity from dilute solutions of mixtures of simple molecules in a hot, reducing ocean under a high pressure of $CO_2$ that I don't know how to connect the two. We need a really good new idea. That idea would, of course, start us down the path toward systems that evolve autonomously—a revolution indeed.

The speech by Whitesides illustrates what Humanist scientists admit to each other and the problems with evolution that are discussed openly in the technical journals. However, the Humanist educational machine continues teaching children that evolution is settled science. See also
http://www.uncommondescent.com/intelligent-design/a-world-famous-chemist-tells-the-truth-theres-no-scientist-alive-today-who-understands-macroevolution/

**Source of the Information is the Achilles Heel**

The effort that evolutionists make to fog up the question of life's origins, such as the PBS story "In Search of Life's Origins" reveals the well-known impasse in origin-of-life studies. In the Prologue to his 2013 *NYT* bestseller, *Darwin's Doubt*, Stephen Meyer explained:

> The type of information present in living cells—that is, "specified" information in which the sequence of characters matters to the function of the sequence as a whole—has generated an acute mystery. No undirected physical or chemical process has demonstrated the capacity to produce specified information starting "from purely physical or chemical" precursors. For this reason chemical evolution theories have failed to solve the mystery of the origin of the first life—a claim that few mainstream evolutionary theorists now dispute.

Why Dr. Meyer wrote *Darwin's Doubt* is an interesting story in itself that he explained in its Prologue. He wrote that in his 2009 book, *Signature of the Cell*, he reported on the impasse (no explanation for the origin of first life) and argued the case for intelligent design. Although that book was limited to the origin of *first* life and the inadequacies of theories of chemical evolution that attempt to explain it, the book received a surprising response. Meyer said that "most criticized the book as if I had presented a critique of the standard neo- Darwinian theories of *biological* evolution." He found that most of his critics sought to refute his claim that no chemical evolutionary process had demonstrated the power to explain the *ultimate* origin of information in DNA (or RNA) necessary to produce life from simpler preexisting chemicals in the first place by citing processes at work in *already living organisms*. In other words, the responders touted the supposed process of natural selection acting on random mutations in *already existing sections of information-rich DNA.* The critics proposed an undirected process that acts on preexisting information-rich DNA to refute Meyer's point that undirected material processes could not produce the information in DNA in the first place.

Perhaps there were so many phony arguments against Meyer's straightforward and accurate claim because Humanists recognized that Meyer's logical conclusion for the ultimate cause of first life-- intelligence -- could unravel the whole evolutionary story. Meyer explained in *Darwin's Doubt* that he had long doubted that mutation and natural selection could add enough new information of the right kind to produce the large-scale changes supposed to have happened even after the origin of life in some form. However, for the sake of argument he had conceded that possibility. He said he found it "increasingly tedious" to concede the substance of arguments he thought were "likely to be false." The evolutionists criticism of arguments he did *not* make in *Signature of the Cell* motivated him to write the present book that responds to the supposed undirected evolution of living things from a common ancestor which is the story conveyed by textbooks, the popular media and spokespersons for "official science." For more on the impasse in origin of life research see http://creation.com/origin-of-life-questions .

### What Was Darwin's Doubt?

The book takes its title, *Darwin's Doubt*, from something Darwin expressed in his famous work, *The Origin of Species.* Darwin was unable to explain in the light of his theories the fossil record which documented the sudden appearance of so many new and "anatomically sophisticated" creatures in the sedimentary layer called the Cambrian without any evidence of simpler ancestral forms. In other words, there were no ancestors and no transitional fossils in that sedimentary layer before the one called the Cambrian layer by those who believe the earth went through geologic epochs of millions of years. Darwin:

> The difficulty of understanding the absence of vast piles of fossiliferous strata, which on my theory were no doubt somewhere accumulated before the [Cambrian] epoch is very great ...I allude to the manner in which the numbers of species of the same group suddenly appear in the lowest known fossiliferous rocks."

The sudden appearance of fully formed animals with no ancestors and no intermediate forms did not accord with his theory of gradual change. *Darwin's Doubt: The Explosion of Animal Life and the Case for Intelligent Design* explores every theory of biological evolution by an extensive review of the books and papers published by evolutionary biologists. Meyer quotes them and shows why their theories can't explain the source of the information contained in living things. It is an enjoyable read. It is loaded with the latest natural science discoveries of which even serious students of the evolution controversy may not be aware. This is the book for any reader of my book who wants a really good technical explanation of all of the improbabilities of evolution that modern science has exposed. As one reviewer of Meyer's book noted:

> Darwinists keep two sets of books. The first set is the real record within peer-reviewed literature that discusses why the mechanism of the origin of life and the mode and tempo of speciation are more baffling today than they were two centuries ago. The second set of books is the popular literature that promotes to the public a soothing, fanciful narrative claiming that the grand history of life is fully explained with only minor but exciting details left to be filled in. Stephen Meyer [audits] the second set of books using the data found in the first.

### Lipstick on a Corpse

The effort that was put into the PBS Evolution Project and the companion book *evolution: The Triumph Of An Idea* in 2000 reflects the desperate crisis in that 19$^{th}$ Century theory — almost like the last shout of a drowning man or putting lipstick on a corpse. New discoveries by experimental scientists, not only in genetics but in other areas as well, have brought into question the evolutionary dogma. There is enormous disparity between popular representations of the status of the theory and its actual status as indicated in peer-reviewed technical journals. Dr. Meyer noted that

> Evolutionary biologists will acknowledge problems to each other in scientific settings that they will deny or minimize in public, lest they aid and abet the dreaded "creationists" and others they see as advancing the cause of unreason...It is an understandable, if ironic, human reaction, of course, but one that in the end deprives the public of access to what scientists actually know. It also

perpetuates the impression of evolutionary biology as a science that has settled all the important questions at just the time when many new and exciting questions---about the origin of animal form, for example---are coming to the fore.

## The Intelligent Design Movement

Biochemist Michael Denton, M.D., Ph.D., published *Evolution: A Theory in Crisis* in 1985. In it he presented a systematic critique of Neo-Darwinism ranging from paleontology, fossils, homology, molecular biology, genetics, and biochemistry. Dr. Denton made this prediction in that book:

> It would be an illusion to think that what we are aware of at present is any more than a fraction of the full extent of biological design. In practically every field of fundamental biological research ever-increasing levels of design and complexity are being revealed at an ever-increasing rate. The credibility of natural selection is weakened, therefore, not only by the perfection we have already glimpsed but by the expectation of further as yet undreamt depths of ingenuity and complexity.

What Dr. Denton wrote in 1985 has certainly come to pass more than he imagined. Engineers have developed a whole new field called biophotonics, bringing light to the life sciences. As just one example, they have invented super-resolution microscopy that enables life science researchers to observe dynamic biological processes inside living cells with unprecedented clarity.

Dr. Denton described himself as an agnostic who rejects biblical creationism. However, so impressive was his critique that it inspired what is now known as the Intelligent Design (ID) Movement which promotes the opposite view to Darwin's unguided random chance theories. The ID Movement is not informed by Divine Revelation and seems to concede geological naturalism if only for argument's sake. So it is not in harmony with the Catholic doctrine of creation but it definitely contradicts Humanism's doctrine of biological naturalism. Dr Denton, even as a non-Christian outside observer to the creation/evolution debate, understood the centrality of *Genesis*:

> As far as Christianity was concerned, the advent of the theory of evolution and the elimination of traditional teleological thinking was catastrophic. The suggestion that life and man are the result of chance is incompatible with the biblical assertion of their being the direct result of intelligent creative activity. Despite the attempt by liberal theology to disguise the point, the fact is that no biblically derived religion can really be compromised with the fundamental assertion of Darwinian Theory. Chance and design are antithetical concepts, and the decline of religious belief can probably be attributed more to the propagation and advocacy by the intellectual and scientific community of the Darwinian version of evolution than to any other single factor.

Alienation of Catholic youth is taking place "right under our noses" in schools everyday and it takes a non-Christian to point it out

**Faith-Affirming Science**

At the end of *Darwin's Doubt*, after he has explained why intelligent design is a scientific theory superior to the various variations of Darwinism, Stephen Meyer made this final point.

> The theory of intelligent design is not based on religious belief, nor does it provide a proof for the existence of God. But it does have faith-affirming implications precisely because it suggests the design we observe in the natural world is real, just as a traditional theistic view of

> the world would lead us to expect. Of course, that by itself is not a reason to accept the theory. But having accepted it for other reasons, it may be a reason to find it important.

Please take time to watch Dr. Meyer's brilliant expansion of why it "may be a reason to find it important."
https://www.youtube.com/watch?v=dvMQXzidVG4
In *Aquinas and Evolution,* Thomistic scholar Fr. Michael Chaberek observed that abstract philosophical arguments for the existence of God are

> more certain and permanent than scientific ones. Scientific arguments, however, are more concrete and easier to grasp for those who have not possessed the ability of abstract thinking. And this is why the persuasive power of the scientific arguments for ID often turns out to be greater than the philosophical arguments for the existence of God. And this is why ID creates more resistance among unbelievers than any of the five ways [of proving God's existence] proposed by Aquinas.

Our Catholic youth can be taught the faith-affirming facts of natural science to counter the faith-destroying propaganda of the Humanists dogma of evolution. But it is a job that must be organized, led and encouraged at the parish level. Can we get parish priests to facilitate the teaching of faith-affirming natural science?

### The Power of the Evolution Alliance

The evolution alliance comprised of schools, universities, and public and private institutions is too powerful for many parents to combat on their own. The power of the evolution alliance was well-illustrated by an incident in 2004. The editor of a biology journal, a man with Ph. D.'s in evolutionary biology and systems biology, incurred a severe penalty for publishing an article that argued that intelligent design could help explain the origin of biological information. The journal was *Proceedings of the Biological Society of Washington* published by the Smithsonian Institution Museum of Natural History, a Federal Government facility whose employees are in the Federal Civil Service with all of the job protection rules that make it virtually impossible to fire anyone. The article provoked a national controversy. The evolutionist alliance was furious with Richard Sternberg for allowing the article to be peer-reviewed and publishing it. Museum officials removed him from office and transferred him to a hostile supervisor. They tried to get him to resign but when that failed, they demoted him. Yet, the offending article itself drew no rebuttal because of the typical dodge evolutionists use to avoid debate: they didn't want to dignify it by responding. Humanists control the popular media, the science journals, the research grants, the universities and the public schools. Teachers and grant-dependent research scientists who want to remain employed must follow the party line. The lesson taught to Richard Sternberg was not lost on them.

### Science Education Powerhouse

The ID Movement is led by the Discovery Institute's Center for Science and Culture founded in 1990 and based in Seattle. It has become another science education powerhouse. Based on its mission statement one can see that it is something every Catholic could support if for no other reason than its support for academic freedom and free speech.

> The mission of Discovery Institute's Center for Science and Culture is to advance the understanding that human beings and nature are the result of intelligent design rather than a blind and undirected process. We seek long-term scientific and cultural change through cutting-edge scientific research and scholarship; education and training of young leaders; communication to the general public;

and advocacy of academic freedom and free speech for scientists, teachers, and students.

The Discovery Institute is highly active in many areas. Check out its website. http://www.discovery.org/id/

## Anti-Intelligent Design Jesuit Nonsense

An article published in November 2005 in a UK newspaper, *The Register*, reported the ramblings of a Jesuit who was the Vatican's chief astronomer. According to the Rev. George Coyne Intelligent Design is not science, and has no place in science lessons. Speaking informally at a conference in Florence he said that intelligent design "isn't science, even though it pretends to be." He argued that if it is to be taught in schools, then it should be taught in religion or cultural history classes, but that it should not be on the science curriculum. Father Coyne has consistently argued against regarding intelligent design as scientific. In June 2005 he wrote in the Catholic magazine *The Tablet*: "If they respect the results of modern science, and indeed the best of modern biblical research, religious believers must move away from the notion of a dictator God or a designer God, a Newtonian God who made the universe as a watch that ticks along regularly. "God," he wrote, is not "continually intervening, but rather allows, participates, loves."

## ID and the Cardinal

Christoph Schönborn, Archbishop of Vienna, had an Op-Ed in the *NY Times* on July 7, 2005 defending ID. He wrote that "evolution in the neo-Darwinian sense -- an unguided, unplanned process of random variation and natural selection", that is, atheistic evolution, "is not true." In his book *Chance or Purpose?* (Ignatius Press, 2007), Schönborn opined that while it is legitimate for those doing research "along strictly scientific methodological lines" to exclude the search for purpose, or finality, from their way of studying nature, it is illegitimate and, indeed, irrational for them to conclude from their findings that there is no purpose, or finality, in the world of nature. And so he reasoned that the aggressive manner in which many working scientists have opposed the group of American scientists who are searching for more evidences of intelligent design in the natural world "does not have much to do with science."

In his review of Schönborn's book, Msgr. John F. McCarthy, J.C.D., S.T.D. wrote that the book is an effort to show how the biological evolution of the human body fits in with the divine plan for man. "The Cardinal clearly presents the viewpoint of a theistic evolutionist, but it is odd that he accepts so confidently Darwin's claim of the descent of man from lower animals while, at the same time, admitting that, concerning the validity of the theory of evolution, 'so many questions still remain open' (p. 168)." [For example, those identified by Pius XII and science.]

**Theistic Evolutionists Are Vague About Details**

Schönborn's opinion left open to his admirers the possibility that, with God's guidance, evolution happened essentially as the neo-Darwinists say, through mutations and natural selection. But, is it really credible to say that all of the specified information needed to develop all of the different plants and animals was contained within the original goo (or whatever) created by God? As Stephen Meyer showed in *Darwin's Doubt*, mutation and natural selection could not add enough new specified information of the right kind to produce the large-scale changes supposed to have happened even after the origin of life in some form. Further, as related on page 49, empirical scientists such as John Sanford and Daniel Lieberman testify that mutations remove information from the human genome and are harmful. Theistic and atheistic evolution are functionally identical. Both are based on 19th century and tooth-fairy science. The only distinction is the empty theological language attached in the former case.

# Chapter 6-The Importance of Noah

This writer understands that many theistic evolutionists believe literally in a world-wide flood as described in *Genesis*. In this chapter I will explain that evolutionary theory is incompatible with any sort of a world-wide flood as described in *Genesis*. Evolutionary theory requires a very ancient earth. The theory that the earth is very ancient depends on uniformitarian geology. Uniformitarian geology rejects Noah's Flood.

Humanist philosophy harkens back to the Greek philosophers who lived hundreds of years before Christ, and many of those believed in evolution and a very old Earth. Therefore, Noah's flood had always been a stumbling block to the spread of Humanist philosophy for a number of reasons. For example, it was a supernatural event and it tended to be a marker in the *Genesis* chronology that gave an approximate age to the Earth.

Also, the *Genesis* flood explains the obvious "wear and tear" on the Earth's crust and the enormous amount of deeply-buried organic material found on every continent and below all of the world's oceans. If there was no such flood, then the "wear and tear" was caused by presently observed natural causes and for such natural causes to have done that much "wear and tear" would have taken billions of years according to uniformitarian geology. How uniformitarian geology became the scientific consensus and paved the way for Darwinian evolution theory is explained in this chapter.

### What Darwin and His Peers Did Not Know

According to the Humanist interpretation of history, the 18th Century was the "Age of the Enlightenment" and coincided with the "Scientific Revolution." Much good empirical science and bad speculative science was done, but the quality of 18th Century natural science was exaggerated by Humanist philosophers of the period. In *Darwin's Black Box: The Biochemical Challenge to Evolution* Dr. Michael J. Behe explained that scientists of that and later centuries thought they understood more than they really did because when they looked at things, it was like looking at a closed black box. They could observe input and output of systems but had no idea regarding the "irreducible complexity" inside the box that modern science has discovered. Behe's book helped to start the Intelligent Design Movement.

**History Invented as the Bible is Jettisoned**

One aim of the Enlightenment was to challenge the authority of beliefs that were deeply rooted in society, such as those inherited from Christian culture. Enlightenment thinkers wanted moral autonomy. Philosophers speculated on ways to reform society with toleration, science, and skepticism. In that atmosphere, alongside of true empirical science, there developed a new genre of science, "pre-historical science." These new pre-historical sciences were called science because they included observation, data collection, naturalistic hypotheses regarding the data, and logical reasoning. Inferences could be drawn, but conclusions could not be proved or disproved, and they lacked predictive value. Humanists championed the acceptance and propagation of pre-historic science hypotheses that helped them explain away Noah's flood and to propose a new "history" of the world. Stephen Jay Gould pointed out in his 2002 book the ongoing problem that evolutionists continue to have because of "canonical stories" that disagree with the history of the world according to practitioners of historic science. ("Canonical stories" is his snarky term for the Bible.) Professor Gould seems to have expressed the same wish as the 18th Century Humanists, namely, that belief in the Bible would just go away.

Humanists embraced and propagated pre-historic science hypotheses that were compatible with naturalism. Scotsman James Hutton, in 1788, published his *Theory of the Earth.* Except for two years studying 18th Century medicine in France and the Netherlands, his biography doesn't say that he ever left Scotland; yet he proposed a theory that encompassed the world. From touring around Scotland, he concluded that the rocks forming the Earth were formed in fire by volcanic activity, with a continuing gradual process of weathering and erosion wearing away rocks, which were then deposited on the sea bed, re-formed into layers of sedimentary rock by heat and pressure, and raised again. For the features on the Earth to have formed in that manner required a long, long time and so began the theory of geologic or deep time. That is how the atheists' modern theory that the Earth is ancient, billions of years old, started. It was not the *evidence* that set the age at billions of years, but rather the naturalistic *interpretation* of the observations of data. The interpretation that features of the earth require a long, long time to form was "exploded" by the 1980 explosion of Mt. St. Helens. https://creation.com/lessons-from-mount-st-helens

Hutton was no "pure scientist." He was a part of the "Scottish Enlightenment" and associated with its famous Humanist philosophers, such as David Hume. Hume wrote *A Treatise of Human Nature* (1739) in which he outlined a totally naturalistic "science of man" that examined the psychological basis of human nature. The phrase "If it feels good, do it" did not originate in the 1960s Cultural Revolution in America. Hume held that ethics are based on feelings rather than abstract moral principles.

Hutton's theory was slow to catch on, but it was embraced by Humanists because it did away with Noah's flood, which Christians thought was a better explanation for the "wear and tear" on the Earth's crust. Hutton's theory provided the vast ages of time that evolutionists invest with creative power. Hutton's theory is known as Uniformitarianism, although the name was coined by another. The theory assumes that the same natural laws and processes that operate in the universe now have always operated in the universe in the past and apply everywhere in the universe. It has included the gradualistic concept that "the present is the key to the past," and is functioning at the same rates. According to his hypothesis, the history of the Earth could be determined by understanding how processes such as volcanism, erosion and sedimentation work in the present day.

In the next century, Charles Lyell, a trained lawyer, was studying geology and heard lectures by the Protestant clergyman and geologist William Buckland who, at that point in his career, was opposing Hutton's theories. In 1820 Buckland published *Connexion of Geology with Religion explained,* both justifying the new science of geology and reconciling geological evidence with the biblical accounts of creation and Noah's Flood. This theory was Catastrophism. But Hutton's theories kept gaining acceptance in the university philosophy circles so Buckland developed a new theory in which he tried to have it both ways, that is, to accommodate the geologic deep time proposed by Hutton and to stick with the Bible.

**Lyell and Darwin**

Charles Lyell would have none of that. He took up Hutton's ideas, expanded them, and popularized his theoretical framework in an 1830 book, *Principles of Geology: being an attempt to explain the former changes to the Earth's surface, by reference to causes now in operation.* This hypothesis, the opposite of the then-consensus belief in Catastrophism, caught on immediately in the swelling ranks of Humanism because it also explained away a supernatural event like Noah's flood and provided a naturalistic hypothesis to assign a vast age to the Earth. See http://creation.com/the-science-of-charles-lyell

Nearly everything Lyell speculated about was accepted by Humanist opinion makers because, as Darwin recognized before he published his famous book in 1859, evolutionary theories of biology depend on timescales so long that almost anything can be proposed without possibility of disproof. Paradoxically though, evolution cannot be proved by any "causes now in operation," which is the working principle of evolutionary geology. Charles Lyell is the one who arranged for divinity student and amateur biologist Charles Darwin's passage on the British Navy ship that took him to the Western Hemisphere, where he developed the basics of his theory. Darwin took Lyell's book with him and made observations through the lens of Lyle's vast ages.

## The Debate Hinges on This

Determining whether Noah's flood took place as described in the Bible or not is the key to understanding the whole debate about evolution. It perhaps explains why *The Genesis Flood* (published 1961) had such an impact on those who read it. It included observed data about the earth that only make sense when explained by a cataclysmic world-wide flood. Among other things, *The Genesis Flood* documented plenty of scientific data that indicate we live on a young Earth and not one that is 4.5 billion years old. Evolutionists never mention this data and they do not mention the assumptions upon which their "ancient earth" estimates depend. As one example, consider the magnetic field of the earth. The magnetic field is a shield essential for our survival because it deflects destructive cosmic radiation from space around the Earth. It has been measured repeatedly and the data, when plotted on a graph, show the strength of the field has decreased and is decaying exponentially. If as uniformitarian geology maintains, the present is the key to the past, then that data can be extrapolated backwards. Extrapolating 10,000 years backward on that exponential curve indicates the magnetic field would be so strong that the Earth would be uninhabitable. That would indicate the Earth is young as the Bible teaches. The solution to that problem offered by evolutionists, because they "know" the Earth is billions of years old, is that the magnetic field reversed itself many times over the course of Earth's history. How, when and why these reversals happened has remained in the realm of naturalistic speculation.

### New Data Shake Up 100-Year Old Hypothesis

According to an article in the July-August 2014 issue of *Discover* magazine, "Journeys to the Center of the Earth," scientists claim to know that Earth's magnetic field is caused by the slow, convective, sloshing of liquid iron in the Earth's outer core (estimated to be 1800 miles below the surface), aided by Earth's rotation." That is the "geodynamo hypothesis" that originated in 1919 but proving it is as yet beyond the reach of any theory, computer simulation, or experiment in the nearly 100 years since. According to the *Discover* article, "[E]vidence from ancient rocks reveals that Earth's geodynamo has been running for at least 3.5 billion years." The *Discover* article continued:

> Two years ago, a team of scientists from two British universities discovered that liquid iron, at the temperatures and pressures found[according to theory] in the outer core, conducts far more heat into the mantle than anyone had thought possible…This discovery is vexing. If liquid iron conducts heat into the mantle at such a high rate, there wouldn't be enough heat left in the outer core to churn its ocean of iron liquid. In other words, there would be no heat-driven convection in the outer core. 'This is a big problem," says Alfe [the lead researcher] "because convection is what drives the geodynamo. We would not have a geodynamo without convection."

The article went on to say that a team in Japan validated the British team's result regarding the heat transfer properties of iron at high pressure.

> Based on how fast Earth's core is cooling and solidifying now, it's likely that the inner core formed relatively recently, perhaps within the past billion years. How did the geodynamo manage to function for at least a couple of billion years before the inner core existed? "The problem is actually in Earth's past," not in the present says Alfe. "This is where new hypotheses are coming in. Some people are saying maybe Earth was a lot hotter in the past."

Note for Mr. Alfe: Have you ever considered that Earth is younger than evolution-based rock dating indicates? Look here https://creation.com/do-creationists-cherry-pick-discordant-dates

The rest of the article is filled with speculations about how the Earth became hot in the first place that "would cross a rabbi's eyes" and ought to teach anyone who reads it with a grain of skepticism what a farce evolution-based science has become. For a more thorough discussion of Mr. Alfe's research use search tool at ICR.org and find "Earth's Young Magnetic Field," Feb. 2016. For more on the "Dynamo Hypothesis" see http://creation.com/moons-magnetic-puzzle

## The Over Simplified Tectonic Plate Theory

Another geologic "fact" that every school child learns is that the various continents are the result of land mass shifts caused by the movement of tectonic plates over millions of years. Following early, unsuccessful theories of continental drift, such as that of Alfred Wegener in the early 1900s, plate tectonics (PT) was introduced in the early 1960s and was quickly adopted by most geologists. Despite widespread acceptance, it remains essentially unchanged and continues to include nagging, unresolved problems. Mark McGuire, a registered professional civil engineer who has poured through the scientific literature has noted that:

> On the surface, plate tectonics is a simple and elegant model that explains many features of Earth's geology. A closer look reveals a number of inconsistencies. These can be found in several key areas, including the number of boundaries of the plates, plate mechanics, mechanisms of plate motion, and the nature of the famous sea-floor magnetic stripes. In its transition from model to paradigm, plate tectonics has lost internal mechanisms to distinguish data from interpretation and to evaluate other potential explanations. [Read http://creation.com/inconsistencies-in-the-plate-tectonics-model]

Maybe the cataclysmic Noah's Flood explains the Earth's geology better than the popular theories of science. God knows.

**Fossil Fuel**

Fossil fuel is a general term for buried combustible geologic deposits of organic (derived from living matter) materials that have been converted to crude oil, coal, natural gas, or heavy oils by exposure to heat and pressure in the earth's crust. According to evolutionists this happened over hundreds of millions of years. When Hutton and Lyell were sitting in Great Britain centuries ago and writing the geologic history of the world under candles or a lamp burning whale oil, did they ever imagine that vast reservoirs of crude oil, natural gas, and coal would be found from Pole to Pole on land and under the oceans thousands of feet below the surface? How, one wonders, would they account for the massive and deep burial of the organic material? In Wyoming there are coal seams 200 feet high. Oil drilling rigs have drilled wells 9,500 feet into the ocean floor. Since the mid-19th Century Humanists and other proponents of evolutionary geology have been repeating that "the present is the key to the past." Earthquakes, volcanoes, sedimentation and erosion of the type happening in recent history did not clump those trillions and trillions of tons of plants, trees, marine algae, and animals together and rapidly bury them below the surface in pockets and seams found everywhere on earth. And it doesn't take millions of years to make fossil fuel. Crude oil can be made from marine algae in an hour. See http://creation.com/algae-to-oil Crude oil is primarily from rapidly buried marine algae. The only explanation that makes any sense is Noah's Flood.

Those with superficial knowledge of *Genesis* 6-8 say "Oh, I know, it rained for 40 days and nights." It is impossible for the system that now produces rain to produce rain, heavy rain, for 40 days and nights continuously to flood the earth. What *Genesis* 7:11 says is that "all the fountains of the great deep burst forth and the windows of the heavens were opened." Most of the water came from underground. Even today there is more water in the earth's mantle than in its oceans. It was a cataclysmic event.

Jesus said to His unbelievers "If you believed Moses you would believe me for he wrote about Me."(John 5:46). This writer suggests that 'if you believed Moses you would believe in Noah's flood because Moses wrote about it. If one believes Moses, logically he can't believe Humanists Hutton and Lyell.

## Arctic Fossils

*The Genesis Flood* has been mentioned in this book as a source for all sort of evidence for a world-wide catastrophic flood as described by Moses. Back in 1961 that book described how buried in the Arctic regions there is vast evidence that before the Flood that area of the world was no colder than elsewhere because it contains so many fossils of animals and flora now found only in temperate and tropical regions. The evolutionists also know the Arctic was once warm and wonder why because they reject the Flood. The April 2015 issue of *Discover*, an evolution science magazine, contained an article titled "Cold Case: Is our climate's future written in Arctic fossils from a warmer past?" The article is about a Canadian paleobiologist named Rybczynski "looking for clues about past global warming" in a period, according to evolutionary dating, "3 to 3.3 million years ago." This contradicts the current Humanist claim that humans caused the current global warming, which so far has been only 0.7°C since 1880.

> Today Ellesmere [Island], which lies next to Greenland on the eastern edge of Canada's Arctic Archipelago, supports only ankle-high tufts of cotton grass and mossy ground cover; the nearest tree is almost 1000 miles south. But Rybczynski and her colleagues have unearthed evidence of a balmier Arctic from a time referred to as the mid-Pliocene warm period, roughly 3 million to 3.3 million years ago. The Island's treasure trove of fossils, preserved in permafrost suggests the area was once an ancient boreal-like forest of larch cedar and birch grazed by miniature beavers, three-toed horses and black bear ancestors.

The article went on to explain that Rybczynski had once found a bone in that Arctic wasteland that was later identified as the tibia bone of a camel that was 30% larger than that of a modern camel. Ironically, the identification was made possible because of collagen in the bone. Collagen is a tough structural protein that ties or connects other tissues such as skin and bones. Since this material should have completely decomposed after only thousands of years, none should be left after the millions of years assigned to these remains. An evolutionist colleague of Rybczynski named Ballantyne was puzzled by these findings regarding the forests that once were there.

> For a productive forest to grow, Ballantyne explains, temperatures have to remain above freezing for half the year...Three distinct data sets pointed to the same number: an average yearly temperature 34 F warmer than today's Arctic...While average global temperatures in the mid-Pliocene rose only 3.6 to 5.4F, the Arctic was a totally different world. "So the question is what was amplifying temperatures in the Arctic?" Ballantyne asks.

*Smithsonian*, April 2016, included testimony about Arctic fossils from the evolutionist director of the National Museum of Natural History. Kirk Johnson spent two summers on Ellesmere Island. "We were finding fossils from a much warmer world. We found a rhinoceros skeleton on an unnamed river, so we named it—it's the only Rhinoceros River in Canada. We found petrified forests and crocodiles and turtles and early mammals."

Why was the Arctic warm at one time and supported forests and animals where there is now a desert of ice? One hypothesis is based on *Genesis* 1: 6-7:

> And God said, 'Let there be firmament in the midst of the waters, and let it separate the waters from the waters. And God made the firmament and separated the waters which were under the firmament from the waters that were above the firmament." And it was so.

Translation of the Hebrew word *raqiya* went to Greek to Latin to English to become "firmament" but the latest research translates it as "expanse," that is, the space around the Earth. Picture the Earth as we now know it, surrounded by air which becomes thinner with the height above the Earth. The main protection for the Earth from the harmful cosmic radiation from the sun and other bodies in space is the magnetic field and the ozone layer. But, if at the Creation, beyond that air, instead of the empty, airless vacuum of space there was a canopy of water mist, what would be the effect? The answer is a green house effect. The rays of heat-bearing light coming toward Earth would hit that water mist and be deflected around the Earth so that instead of the Equatorial Region being hot and the North and South Poles being cold, the whole Earth would be temperate. The Earth's mild climate would enable the growth everywhere of the lush vegetation necessary to support large populations of people, plants and animals, even for vast consumers like dinosaurs.

If there was a water mist canopy as that hypothesis proposes, what happened to it? *Genesis* 7:11 says that "all the fountains of the great deep burst forth and the windows of the heavens were opened." What were those "windows of heaven" that opened? There is no mist up there now. I am not suggesting that the mist (if it was there) was the source of the 40 days of rain and a major cause of the flooding. Scientific modeling has ruled out that amount of water in the theoretical canopy because the "greenhouse" would have been too hot. But God could have "fine tuned" the amount of mist and its distance from Earth to fulfill His purpose. It's only a hypothesis that could never be proved but it may explain why the Arctic was once warm.

If the water for the Flood did not come down, did it come up? The Hydroplate Theory of Walter Brown is one possible explanation. When "all the fountains of the great deep burst forth," water above the crust of the Earth and much more below created massive fountains spurting miles into the sky with enormous pressure to create the "rain." There is a crack that goes all of the way around the Earth under the oceans and looks like the seams on a baseball. The seafloor is dotted with thousands of steep-sided underwater volcanoes, or seamounts. See Appendix III for Brown's book *In the Beginning: Compelling Evidence for Creation and the Flood.* In 2004 a tsunami from just one

underwater earthquake in the Indian Ocean killed 250,000 people. Watch and try to picture when "all the fountains of the great deep burst forth."
youtube.com/watch?v=yjtKRRPhWPA
The scaring of the Earth, its jagged mountains with marine fossils on top of them, the wasteland deserts, the trillions and trillions of tons of deeply buried plants and animals that became fossil fuel, the massive fossil grave yards and the ice age all point to a catastrophic Flood. According to evolutionists at the Smithsonian Institution:

> Sixty-five million years ago the dinosaurs died out along with more than 50% of other life forms on the planet. This mass extinction is so dramatic that for many years it was used to mark the boundary between the Cretaceous Period, when the last dinosaurs lived, and the Tertiary Period, when no dinosaurs remained.

You may have learned in school that the cause of the dinosaurs' extinction was an asteroid impact in Mexico. That's bogus science. Read http://www.icr.org/article/chicxulub-crater-theory-mostly-smoke/. The "mass extinction" is consistent with a world-wide flood. Read http://www.icr.org/article/9748/. After the Flood, dinosaurs and other animals that came off of Noah's Ark would have been competing with man for food and had difficulty foraging on the recently denuded planet. The Cretaceous and Tertiary Periods are just different layers of water- deposited material that hardened into rock, that is, sedimentary layers. Based on the Hutton-Lyle Theory, the layers were deposited millions of years apart. But deposits of sedimentary layers can be laid in hours.

What do marine fossils on the top of Mount Everest tell about the Biblical Flood? Fossils of marine creatures in limestone near the summit mean that this area must have been under the sea in the past. Many people would not associate the limestone and fossils with Noah's Flood because they think there was not enough water to cover the highest mountains. However, according to *Psalm 104:6-9*, the Flood changed the earth's topography. The mountains rose and the valleys sank down at the end of the Flood. With vertical earth movements towards the end of the Flood, the mountains rose and the water flowed off the continents into the newly formed oceans' basins. That's one explanation for why there are marine fossils at the tops of high mountains.

**The Pope Who Authoritatively Taught about Noah's Flood**

If this writer has not succeeded in communicating to theistic evolutionists who do believe literally in Noah's Flood that such a Flood negates uniformitarian geology which, in turn, negates the billions of years during which evolution is supposed to have happened, he does not know what else he can say. At this point the writer turns to Catholics who believe what the Church teaches but do not believe literally in the world-wide flood described in *Genesis*. Catholics may not have heard that there was a Pope who, for our belief, taught about Noah's flood according to *Genesis*. He predicted that "scoffers" will "deliberately ignore this fact that by the word of God heavens existed long ago, and an earth formed out of water and by means of water, through which the world that then existed was deluged with water and perished." Not only that, this Pope taught that when that ancient world perished because God brought a flood upon it, He waited patiently while Noah built the ark in which eight persons were saved. That Pope was Peter. Read it in 1 Peter 3:20, 2 Peter 2:5 and 2 Peter 3:5-6. Why is it that Evangelicals put more belief in our Pope's teaching about the *Genesis* flood than many Catholic scholar-priests and lay intellectuals do?

If one doubts the first Pope will one believe it from the lips of the Second Person of the Trinity? Read it in Matthew 24:28.

> For as in those days before the flood they were eating and drinking, marrying and giving in marriage, until the day when Noah entered the ark, and they did not know until the flood came and swept them all away, so will be the coming of the Son of man.

One who denies the scope of Noah's Flood might think that Jesus must have been talking about a local flood. If that is so, does he judge that Our Lord made a stupid comparison by equating the impact of a local flood with the world-wide impact of His Second Coming?

### You and the Wives of Noah's Three Sons

Earlier in this book the importance and function of DNA was briefly discussed. One type of DNA is passed from generation to generation only through mothers. This genetic material is known as mitochondrial DNA or mtDNA. Typically, a sperm carries mitochondria in its tail as an energy source for its long journey to the egg. When the sperm attaches to the egg during fertilization, the tail falls off. Consequently, the only mitochondria the new human gets are from the egg its mother provided. Ancestry testing has become popular based on new knowledge of genetics. Ancestry.com heavily advertises on TV for people to "Discover your family history and start your family tree." There are three types of genealogical DNA tests. One of those, mtDNA, tests a man or woman along his/her direct maternal line. "DNA Trends Confirm Noah's Family", *Acts & Facts*, July 2016, told how research biologist Dr. Nathanial Jeanson plotted hundreds of human mtDNA sequences and the project revealed an obvious pattern: The mtDNA stemmed from three central "trunks" or nodes instead of just one commonly known as mitochondrial Eve. Jeanson's data suggests that the wives of Noah's three sons best explain that finding. http://www.icr.org/article/new-dna-study-confirms-noah/

Some Catholics mock Evangelicals with the term "Fundamentalists" for taking the Creation and Flood accounts in *Genesis* literally. Many of those same Catholics take Jesus' teaching on divorce literally. (And sadly, many don't.) According to Mark 10:6-8, Jesus said: "But from the beginning of creation 'God made them male and female.' 'For this reason a man shall leave his father and mother and be joined to his wife, and the two shall become one.' So they are no longer two but one." Jesus was quoting from chapters 1 and 2 of *Genesis*. These are the same chapters that theistic evolutionists say can't be taken literally. Why would Jesus quote literally from those chapters to give His definitive teaching on divorce if He didn't intend for us to believe them literally? And notice that He said that they were male and female from the beginning of creation, not after millions of years of evolution. *The Catechism of the Catholic Church*, in its teaching about marriage in paragraphs 1604-1607, quotes from Genesis 1 and 2 seven times. *The Catechism*, in its teaching regarding our Sunday obligation quotes Scripture in paragraph 2169 as follows: "For in six days the Lord made heaven and earth, the sea, and all that is in them, and rested on the seventh day; therefore the Lord blessed the Sabbath day and hollowed it." Catholic theistic evolutionists accept the Sunday obligation but believe in "creation" over billions of years. Go figure. Evangelical fiat creationists divide into "young earthers" and "old earthers. Watch a panel discussing that difference of opinion. https://www.youtube.com/watch?v=5fQMfwHFs6Q
Then watch an Evangelical arguing the case for a young earth. https://www.youtube.com/watch?v=ggJZz3WkTCI

Because I believe the Bible is inerrant I've never fretted over how Noah's Ark contained the kinds of birds and animals that repopulated the Earth. Readers troubled by that question should see http://creation.com/how-did-all-the-animals-fit-on-noahs-ark For more on the Flood see http://creation.com/local-flood

## Our Total Evolution Age

The Enlightenment was more about philosophy than science but at the end of the 19th Century much excitement was generated by scientific theories set forth in Darwin's *Origin of Species* and *The Descent of Man* and similar writings. One can understand how Humanists and Christians alike got caught up in the novelty of it all. But, as time has gone on, all of the empirical science has gone against evolution as the "black boxes" have been opened, as Dr. Behe's book explained. Yet, despite no new scientific support, evolution has gone from a biological theory to the Modern Synthetic Theory of Evolution, a viewpoint that is so broadly applied that one can speak most accurately of Total Evolutionism, including Stellar Evolution, Molecular Evolution, Organic Evolution, and Societal (or cultural) Evolution. Marking the 100th anniversary of *Origin of Species* in 1959, Julian Huxley, the first Secretary General of the UN's Education, Scientific and Cultural Organization (UNESCO), remarked that

> Future historians will perhaps take this Centennial Week as epitomizing an important critical period in the history of this earth of ours—the period when the process of evolution, in the person of inquiring man, began truly to be conscious of itself...this is one of the first public occasions on which it has been frankly faced that all aspects of reality are subject to evolution, from atoms and stars to fish and flowers, from fish and flowers to human societies and values—indeed, that all reality is a single process of evolution.

John N. Moore commented on the social phenomenon of "total evolutionism" in 1977 in *Acts & Facts*. The article was "The Impact of Evolution on Social Sciences":

> However the broadened viewpoint of Total Evolutionism that has developed in a little over a century from Darwinism is without any significant repeatable empirical

data from naturally occurring events. On the contrary evolutionists must speak glowingly and write ingeniously about numerous *supra*-natural concepts; such as, a "big bang" explosion of a dense particle, spontaneous generation of living substance, mountain building due to movement of dry rock masses, division of one land mass into existing continents, and new physical traits through mutational changes. All of these ideas are totally without any empirical support from studies of naturally occurring events of the magnitude involved in such concepts. Yet the circumstantially grounded megaevolutionary point of view involving Total Evolutionism is *the* worldview that has been adopted by influential scholars in every major academic discipline of human thought. Evolution is the *supreme* over-riding point of view or worldview adopted in every major academic discipline by the intelligentsia around the world.

*[How this social phenomenon came to be is explained in chapter 14.]*

### What God Is This?

God has revealed His *fiat* creation in the Bible. Is it consistent with the character of God to reveal an account of His work in creating the heavens and the earth and all they contain, and then to allow all Fathers, Doctors and Popes of His Church to believe and proclaim that account, as written, for almost two thousand years, only to "enlighten" the Church with a radically different, evolutionary account of the origins of man and the universe, not through the work of Catholic saints or scholars but through the speculations of men like Hutton, Lyell, Darwin, Ernst Mayer, and Stephen Jay Gould? And, if it is, who in his right mind would trust such an incompetent, self-contradictory "god"?

# Chapter 7-Evolution and Theology

Jesuit belief in evolution accelerated early in the 20th Century because of the scientific reputation and evolutionary theological writings of another one of their own, Frenchman Pierre Teilhard de Chardin, S.J. Pierre Teilhard de Chardin's greatest contribution to the theological waywardness of the Jesuits was through his writings. Fr. Victor Warkulwiz, M.S.S., credits de Chardin with introducing a new genre of literature, namely, theology fiction. It was mystical evolutionary poetry and prose that, according to Fr. Warkulwiz, mesmerized many Catholics with a "seductive merging of the spiritual and the secular." According to Fr. Warkulwiz, "His writings create havoc with Catholic notions about creation, redemption, sanctification, original and actual sin, evil and grace." In 1957, this theology fiction caused the Vatican's Congregation for the Doctrine of the Faith, then called "the Holy Office," to order that his works be removed from libraries of Catholic institutions and forbade their sale in Catholic bookstores. In 1962, about when Mr. Fitzpatrick was being taught by Jesuits at Fordham, there came another document warning the faithful about errors and ambiguities in de Chardin's writings. Obviously those warnings were, and continue to be, ignored. This writer's wife recalls seeing de Chardin's *The Phenomenon of Man* and *The Divine Milieu* being passed around among the students at her Catholic women's college circa 1962. His books were published and republished. At least nine of his books are for sale online. Somebody is buying them.

### Famous Jesuit Evolutionary Paleontologist

Fr. de Chardin's influence on the Jesuits' theological and philosophical traditions came, in part, from his reputation as a paleontologist. Fr. de Chardin's reputation was greatly enhanced because of his significant role in what was heralded as the greatest paleontological discovery of all time, namely, the Piltdown Man. Piltdown Man is the name given to some fragments of a jawbone and a skull unearthed in close proximity to each other in an English gravel pit by paleontologist Charles Dawson in 1912. Although questioned by some, they were widely regarded to be fossilized bones of an ape partially evolved into a man. In 1913, while digging with Dawson, de Chardin discovered a canine tooth that seemed to provide the confirming evidence. From 1913, these fragments became recognized by the scientific consensus as the "missing link" in the descent of man from a lower animal. It was the proof through fossil evidence that Darwin predicted would be discovered when he published his theory of biological evolution in 1859. Piltdown Man, with plaster filling in the missing parts of the fossil according to the imagination of Dawson the paleontologist credited with the discovery, was prominently exhibited in the British Museum as an example of human evolution until it was discovered to be a complete forgery. The jawbone was that of an orangutan that had been doctored.

It was proved to be a forgery in 1953. That was after generations of school children, perhaps including the above-mentioned Mr. Fitzpatrick, had studied textbooks with artist conception drawings of "Eoanthropus dawsoni," a half man-half ape complete with hunched back, hairy body and a "knowing" look in his eye and had been told that the Piltdown Man was their ancestor. And they believed it.

### Humanist Propaganda Film

Some readers may remember the movie "Inherit the Wind," starring Spencer Tracy. It was a pro-evolution/anti-Bible fictionalization supposedly based on the so-called "Scopes Monkey Trial" which took place in Tennessee in 1925. A high-school teacher, John Scopes, collaborated to create a case for the American Civil Liberties Union (ACLU) to contest Tennessee law. Scopes, who wasn't actually a science teacher, discussed evolution with a student and volunteered to be tried for violating the law which prohibited teaching the evolution of humans. In the real trial, Scopes was defended by a famous defense lawyer and ACLU member named Clarence Darrow who introduced the Piltdown Man as part of the defense. The movie was made in 1960, seven years after the Piltdown Man fake was proved, so naturally Hollywood never mentioned the part that fake had played in the 1925 trial. Gary Parker, Ed. D., in an article in *Acts & Facts*, observed in 1981:

> At least Piltdown answers one often-asked question: "Can virtually all scientists be wrong about such an important matter as human origins?" The answer, most emphatically, is: "Yes, and it wouldn't be the first time." Over 500 doctoral dissertations were done on Piltdown Man, yet all this intense scientific scrutiny failed to expose the fake.

## Vestigial Structures

In addition to Piltdown Man, the Scopes Trial also showcased another belief of the 1925 evolutionist consensus, namely, vestigial structures. The term (in evolution speak) means genetically determined structures or attributes that have apparently lost most or all of their ancestral function in a given species, but have been retained through evolution. This is 19$^{th}$ Century Darwin theory that can be summarized as "use it or lose it." According to Darwin, the effect of "use or disuse" strengthens and enlarges certain body parts. That part is true; just take a walk in Gold's Gym on a day when the heavy lifters are working out. But Darwin claimed that the species have been modified by the inherited effects of the use and disuse of parts. Accordingly, the children of heavy weight lifters should inherit big biceps. Science in Darwin's era preceded the science of genetics.

Darwin wasn't the first believer in evolution. It was a hot topic among Naturalists since the Greek philosophers and it was "in the air" of the 18th Century. Because the rudimentary science of that era did not know what certain body parts were for they assumed they were left over from a lower form from which humans were thought to have evolved. In 1798 a Frenchman noted that

> Whereas useless in this circumstance, these rudiments... have not been eliminated, because Nature never works by rapid jumps, and She always leaves vestiges of an organ, even though it is completely superfluous, if that organ plays an important role in the other species of the same family.

The term "vestigial" was coined later by a 19th Century scientist influenced by Darwin's "use it or lose it" theory. A German, Robert Wiedersheim, in 1893 published a list of 86 human organs that he said had "lost their original physiological significance" and he attributed that to evolution. The 19th Century evolutionists didn't know what God's purpose was for currently observable things in humans so they considered them as leftovers from a previous stage of evolution. By 1925, Wiedersheim's list had grown from 86 to 180. Wiedersheim's list was introduced at the Scopes Trial with the comment that a human had so many vestigial structures left over from evolution that he was a veritable walking museum of antiquities.

### 19th Century Science on Wikipedia

It's almost funny to read the evolutionists commentary on structures when they make up excuses to still regard human body parts as vestigial. Look up the Wikipedia entry for the word "Vestigiality." The appendix was a favorite example of a useless, vestigial structure by "experts" from Darwin to Harvard's Ernst Mayr but then it was discovered to be an important factor in regulating the level of gut microflora that includes major functions such as metabolic activities that result in the salvage of energy and absorbable nutrients, on immune structure and function, and protection against foreign microbes. In the following, notice how the evolutionist Wikipedia writer first presents an evolutionary assumption as a fact before admitting that the appendix is useful.

> A classic example at the level of gross anatomy is the human vermiform appendix — though vestigial in the sense of retaining no significant *digestive* function, the appendix still has immunological roles and is useful in maintaining gut flora.

What evidence is there that the appendix *ever* had a digestive function except the assumption that it did? For more on this major evolutionist blunder see "Our Useful Appendix- Evidence of Design, Not Evolution" in *Acts & Facts*, February 2016, a free magazine from ICR.org. Here is another example

> The coccyx or tailbone, though a vestige of the tail of some primate ancestors, is functional as an anchor for certain pelvic muscles including: the levator ani muscle and the largest gluteal muscle, the gluteus maximus.

That humans had an ancestor is a big enough fairy tale without adding that the ancestor had a tail.

**Isn't Evolution Thoughtful?**

> The emergence of vestigiality occurs by normal evolutionary processes, typically by loss of function of a feature that is no longer subject to positive selection pressures when it loses its value in a changing environment. More urgently the feature may be selected

> against when its function becomes definitely harmful. Typical examples of both types occur in the loss of flying capability in island-dwelling species.

Evolution, "knowing" that the island birds had no need to leave the island, decided to devolve the flying ability it took billions of years for them to evolve from when they were fish or whatever because they might fly out over the ocean and forget how to get home. That would be harmful. Or, because the birds knew they had nowhere to go, they started walking around the island and failing to heed Mr. Darwin's dictum to "use it or lose it," they lost their flying ability.

> Humans also bear some vestigial behaviors and reflexes. The formation of goose bumps in humans under stress is a vestigial reflex. Its function in human ancestors was to raise the body's hair, making the ancestor appear larger and scaring off predators.

Catholic children in public schools have been taught nonsense like this for a long time. Catholic theistic evolutionists were made, not born.

### Whale Sex: It Is In the Hips

For years, evolutionists have pointed to certain bones in whales as vestiges. According to a theory, the whales' ancestor had legs. When that ancestor went to sea, through non use, the legs were lost and those bones are all that is left. A September 8, 2014, release from the University of Southern California disproved that theory. The university published a story on its website titled, "Whale Sex: it's all in the hips." In the article, it announced

> New research turns a long-accepted evolutionary assumption on its head – finding that far from being just

vestigial, whale pelvic bones play a key role in reproduction.

And now that they are shown to be necessary, what evidence is there that they are vestigial of anything? That whales could ever have evolved from land animals is tooth-fairy science. For a great description of the amazing unique design of whales see *Zombie Science* chapter 5 (Discovery Institute, 2017).
For more discussion about vestigial propaganda see http://creation.com/do-any-vestigial-organs-exist-in-humans

**All of the "Links" are Missing**

Darwin theorized that individuals acquired beneficial characteristics and passed them on by natural selection or "survival of the fittest." In this view, evolution is seen as generally smooth and continuous. It required that there be some evidence of "in between" things which are called transitional fossils. That explains the scientific excitement generated by Piltdown Man, the "link" between humans and their supposed non-human ancestors. The fossils known in Darwin's time showed fully-formed individuals although some were subjects of further debate. Some were thought to be of something that had become extinct and some were the same as living things of his era. Darwin explained that transitional fossils hadn't been found because relatively few fossils had been found. According to Darwin, the reason so few had been found was because the earth's crust had been formed by natural processes such as volcanoes, ice movement, wind, rain and erosion. That was the 18$^{th}$ Century uniformitarian theory of geology which, if true, projected the age of the earth to be billions of years. Under those natural conditions, nature would work against a dead organism becoming fossilized. For example, dead animals would be eaten by other animals or birds and their bones would be scattered. If they got buried before they had been picked over, the effects of wind, rain, floods, etc. could uncover them.

When archeology became a much more systematic discipline in the late 19th century and became a widely used tool for historical and anthropological research in the 20th century, more fossils than Darwin ever imagined were unearthed. Millions of fossils have been found. The world's museums are full of fossils. For example, between 1909 and 1915 the Smithsonian Museum collected over 65,000 specimens, many very well preserved, from a site in British Columbia known as the Burgess Shale. These mostly sea creatures are found at 7,500 feet up in the Canadian Rockies. These complex animals had apparently "risen" suddenly, distinct and fully formed, with nothing by way of ancestor forms. Other massive deposits of fossils, distinct and fully formed from the same so-called Cambrian geologic age were discovered in the mid-1980s in southern China. The "Chengjiang Fossils" are an even greater variety, including soft-body animals, than the Burgess Shale.

**Hopeful Monster**

Richard B. Goldschmidt (d.1958) was a famous geneticist. He is considered the first to integrate genetics, development, and evolution. He did important work that advanced the science of genetics. He wrote that at age sixteen

> ...it seemed that all problems of heaven and earth were solved simply and convincingly; there was an answer to every question which troubled the young mind. Evolution was the key to everything and could replace all the beliefs and creeds which one was discarding. There were no creation, no God, no heaven and hell, only evolution and the wonderful law of recapitulation which demonstrated the fact of evolution to the most stubborn believer in creation.

The "law of recapitulation" that teenager Goldschmidt learned is one of the most famous frauds of the evolutionists, Ernst Haeckel's embryos. Among others who believed that German's fraud was the famous German Jesuit theologian Karl Rahner and that was responsible in part for his well-publicized heretical polemics against *Humanae Vitae*. When one thinks he was an animal in his mother's womb, how can he understand what it means to be human? Haeckel's fraud has been very influential for over 100 years. It was repeated in textbooks even after it was well-known to be a fraud. To read how significant Haeckel's fraud was type this in your browser and read about it:
evolutionnews.org/2015/04/haeckels_fraudu094971.html
This fraudulent "science" continues to be perpetuated in textbooks including two titled *Biology* by different authors published in 2014. In a "debate" via Facebook my anti-evolutionism was ridiculed by a Catholic obstetrician who said I obviously had never studied embryology. She probably studied Haeckel's embryos in college and may have used the popular textbook by Douglas Futuyma, *Evolutionary Biology*, in which he wrote that "early in development human embryos are almost indistinguishable from those of fishes."

When Goldschmidt grew up and became a scientist he was uncomfortable with the lack of transitional fossils. He postulated a theory to explain the sudden appearance in the fossil record of fully-formed specimens. Goldschmidt advanced a model of macroevolution[big changes] through macromutations [big mutations] that is popularly known as the "Hopeful Monster" hypothesis In this context, Goldschmidt meant the big jumps that were made by evolution from one fully-formed species to the next "more complex" fully-formed species without intermediate forms were because of really, really big mutations.

## Punctuated Equilibrium

Perhaps the term “Hopeful Monster” was a bit of an embarrassment to evolutionists until that consummate story-spinner and Harvard professor Stephen Jay Gould solved the problem. He refined and changed the name of the theory. With Harvard colleague Niles Eldredge, Gould proposed “punctuated equilibrium.” Eldredge and Gould proposed that the degree of gradualism commonly attributed to Darwin’s theory is virtually nonexistent in the fossil record, and that stasis (no change) dominates the history of most fossil species. In plain speak, there are no transitional fossils. All of the intermediate “links” are not missing, they were never existent. Punctuated equilibrium is a refinement to evolutionary theory. It describes patterns of descent taking place in "fits and starts" separated by long periods of stability. Punctuated Equilibrium sounds better than Hopeful Monster but it’s essentially the same theory. One only has to examine Punctuated Equilibrium to see a typical example of evolutionist fog. This is how Gould and Eldredge explained it in 1977 in *Paleobiology*:

> Punctuated equilibrium proposes that most species will exhibit little net evolutionary change for most of their geological history, remaining in an extended state called *stasis*. When significant evolutionary change occurs, the theory proposes that it is generally restricted to rare and rapid (on a geologic time scale)) events of branching speciation called cladogenesis.

Sounds impressive, right? What is cladogenesis? According to the authors, cladogenesis is the process by which a species splits into two distinct species, rather than one species gradually transforming into another. Get it? The species dog splits into cats and chickens. Why do Harvard professors write such nutty stuff? They know that all of the entirely different species did not “evolve” gradually as Darwin proposed because there are no transitions. Yet they "know" that the various distinct species are the result of evolution. So if species didn't change into different species slowly through small changes, they must have changed into different species rapidly through big changes they had been storing up for a long, long time. Because it is a theory, it qualifies as "science."

### Around in a Circle

In 1980, also in the journal *Paleobiology*, Gould said punctuated equilibrium was a "new and general theory" of evolution and that neo-Darwinism is "effectively dead, despite its persistence as textbook orthodoxy." According to Dr. Stephen Meyer in *Darwin's Doubt* it was "only after critics exposed punctuated equilibrium for lacking an adequate mechanism did Gould retreat to a more conservative formulation of the theory, making its reliance upon the neo-Darwinian mechanism explicit." In other words, according to Meyer,

> advocates of punctuated equilibrium were forced to concede both the inadequacy of their proposed mechanisms and to rely on the neo-Darwinian process of mutation and natural selection in order to account for the origin of new genetic traits and anatomical innovations...Thus, though the theory of punctuated equilibrium was initially presented as a solution to the mysterious and sudden origin of animal forms, upon closer inspection, it failed to offer such a solution.

### No Evolution Proves Evolution

In the Gould-Eldredge punctuated equilibrium fig leaf, "most species will exhibit little net evolutionary change for most of their geological history, remaining in an extended state called *stasis*." Stasis means a period or state of inactivity or equilibrium. Evolutionists even claim that by remaining in an indefinite period of inactivity, that is, by staying exactly the same, evolutionary theory is demonstrated. "By not evolving, deep sea microbes may prove Darwin right" is the title of a 2015 a paper published in *Proceedings of the National Academy of Science*. The paper is about three communities of bacteria, two of which were found fossilized and one is living off the west coast of South America. The first fossils are in a rock which according to evolution-based dating methods is 2.3 billion years old. The second fossils are in a rock evolutionary dating indicates is 1.8 billion years old. All of the samples are identical. According to J. William Schopf, a paleobiologist at UCLA: "In form, function and metabolism, they are identical,"

> Researchers say these microscopic organisms are an example of "extreme evolutionary stasis" and represent the greatest lack of evolution ever seen. They may also, paradoxically, prove that Darwin's theory of evolution is true.

Scientists from the Institute for Creation Research pointed out the obvious: "evolutionary stasis is an oxymoron." When a complete lack of difference is counted as evidence for evolution, and all other differences are attributed to evolution, it shows that evolutions is an arbitrary and unfalsifiable assumption—not even a hypothesis.

## Our Cousins Are Mushrooms

While on the subject of wacky evolutionists, consider this March 2015 online report of an article published in *Genome Biology:* "Evolutionary tree: Humans may have evolved with plant genes, study claims"

> Humans may have evolved with the genes of plants, fungi and micro-organisms, according to a consensus-challenging Cambridge University study. The study into the literal roots of mankind builds on, and to some extent confirms, the findings of a 2001 investigation into

> whether or not humans could have acquired DNA from plants... "We may need to re-evaluate how we think about evolution."

**Evolutionists Discovered Genetics?**

In the 1860s Augustinian monk Gregor Mendel performed experiments that, when recognized and validated nearly 40 years later, provided the basis for the new science of genetics. Darwinian evolution, though still taught to school children, was replaced by a new theory called Neo-Darwinism. Evolutionists define Neo-Darwinism as the "modern synthesis" of Darwinian evolution through natural selection with Mendelian genetics. Neo-Darwinism is the view that evolution is due to the natural selection of variations that originate as gene mutations. (As will be explained below, gene mutations are harmful, not beneficial.) Evolutionists kept Darwin's name alive perhaps to shield the fact that observational genetics of the here and now gave Darwinism a kick in the teeth. In their desperate attempt to "rescue" Darwin some evolutionists have made the absurd claim that Darwinism led to the discovery of genetics. For example, in May 2015 in a blog defending the reliability of a "scientific consensus," theoretical astrophysicist Ethan Siegel wrote:

> Think about evolution, for example. Many people still rally against it, claiming that it's impossible. Yet evolution was the consensus position that led to the discovery of genetics, and genetics itself was the consensus that allowed us to discover DNA, the "code" behind genetics, inherited traits and evolution.

In response to Ethan Siegel, EvolutionNews.org published an excerpt from the *Politically Incorrect Guide to Darwinism and Intelligent Design* that pointed out that Mendel found Darwin "unpersuasive." Darwin believed cells contained what he called "gemmules" that transmit characteristics in a "blending process” Darwin called “pangenesis.” According to Darwin these “gemmules” change by use or disuse. That’s the basis for his “use it or lose it” that evolutionists still believe in when they refer to flightless birds or whales that supposedly began as land animals and lost their legs.

> Mendel's theory of stable factors contradicted Darwin's theory of changeable gemmules. Thus, although Mendel's work was published in 1866, Darwinists totally ignored it for more than three decades. William Bateson, one of the scientists who “rediscovered” Mendelian genetics at the turn of the century, wrote that the cause for this lack of interest was “unquestionably to be found in that neglect of the experimental study of the problem of Species which supervened on the general acceptance of the Darwinian doctrines.”

In other words, the scientific consensus was so enamored by Darwinism that they saw no need for alternative theories until there was so much evidence that Mendel was right. Which raises the question: Does evolution have any scientific value at all?

### Is Evolution of Any Scientific Value at All?

Evolutionist Dr. Marc Kirschner, founding chair of the Department of Systems Biology at Harvard Medical School was quoted in the October 23, 2005 *Boston Globe* as having stated:

> In fact, over the last 100 years, almost all of biology has proceeded independent of evolution, except evolutionary biology itself. Molecular biology, biochemistry, physiology, have not taken evolution into account at all.

In similar vein, the anti-creationist Larry Witham wrote:

> Surprisingly, however, the most notable aspect of natural scientists in assembly is how little they focus on evolution. Its day-to-day irrelevance is a great ‘paradox’

> in biology, according to a *BioEssays* special issue on evolution in 2000. 'While the great majority of biologists would probably agree with Theodosius Dobzhansky's dictum that "Nothing in biology makes sense except in the light of evolution", most can conduct their work quite happily without particular reference to evolutionary ideas', the editor wrote. 'Evolution would appear to be the indispensable unifying idea and, at the same time, a highly superfluous one.' (Witham, Larry A., *Where Darwin Meets the Bible: Creationists and Evolutionists in America* (hardcover), p. 43, Oxford University Press, 2002)

Evolution contributes nothing tangible to science but that doesn't mean that hoary 19th Century notions can't be used to sell books to the gullible public. Cardiologist Lee Goldman dean of the College of Physicians and Surgeons, chief executive of Columbia University Medical Center in a 2015 book aimed at the general public titled *Too Much of a Good Thing,* wrote:

> Can't stick to a diet? That's a holdover from when humans roamed the plains and gorged when food was plentiful, storing the rest as fat for when it wasn't. Anxiety is a descendant of the fight-or-flight response, which kept us alive when faced with a woolly mammoth but is something that we less often need today.

Is that medical science or something he learned in high school? Medical doctors find evolution theory useless because they are more like results-oriented engineers than theoretical scientists. Evolution is a "white elephant"-big and useless.

At http://www.dissentfromdarwin.org read the list of 950 Ph. D. scientists who have signed their name to the following statement:

We are skeptical of claims for the ability of random mutations and natural selection to account for the complexity of life. Careful examination of the evidence for Darwinian theory should be encouraged.

**"Darwin of the 20th Century" Frustrated by Genetics**

Ernst Mayr, an atheist, published *What Evolution Is*. Dr. Mayr was hailed by the *NY Times* as "the Darwin of the 20th Century." At the time of the book's publication in 2001, Dr. Mayr had published 14 books on evolutionary biology and zoology and was Professor Emeritus in the Museum of Comparative Zoology at Harvard University. In the Preface to his book, Dr. Mayr complained that

> ...most treatments of evolution are written in a reductionist manner in which all evolutionary phenomena are reduced to the level of the gene. An attempt is then made to explain the higher-level evolutionary process by "upward" reasoning. This approach invariably fails.

What Dr. Mayr meant by "upward reasoning" is the effort to explain macroevolution based on reasoning "upward" from microevolution. Why that approach "invariably fails" will be explained. To understand the scientific debate it is necessary to understand what each of those terms mean. It will then be understood why observed genetics thwarts the story tellers' fictional fantasies.

A good explanation by Dr. John D. Morris follows. (John D. Morris. 1996. "What Is the Difference Between Macroevolution and Microevolution?" *Acts & Facts*. 25 (10). Copyright © 1996

> *Micro*evolution refers to varieties within a given type. Change happens within a group, but the descendant is clearly of the same type as the ancestor. This might better

be called variation, or adaptation, but the changes are "horizontal" in effect, not "vertical."

All creation-supporting scientists agree that microevolution as described above, “horizontal,” is non-controversial. The disagreement concerns the alleged “vertical” evolution. Darwin’s second major work was *The Descent of Man* and the notion that all the species descended from one ancestor is the modern synthesis of evolution. Thus, the term “vertical” is used to mean the flow of evolution in a vertical direction. Picture a family tree with your two sets of great grandparents at the top and you at the bottom. Your position on the chart as a descendent of your great grandparents would be vertically below them with your other ancestors above you. In your family tree, all of your ancestors would be humans. In an evolutionary family tree, your pre-historic ancestors would be a variety of different and unique animals. (They would be different and unique because of “punctuated equilibrium,” remember.) You may not look at all like your great grandparents, but this is because there are other fully-human genes brought into the family’s gene pool by the spouses of your great grandparents’ descendants, who were not descendants of those great grandparents. The only genes they could bring into the gene pool were their human genes.

**Artificial Selection**

Then there is “artificial” selection. A fairly recent new breed is a Labradoodle. It is a crossbreed of the Labrador Retriever and the Standard, Miniature, or Toy Poodle. The term first appeared in 1955, but was not popularized until 1988, when the mix began to be used as an allergen-free guide dog. Both of the crossbred animals were of the same species. It is possible to crossbreed animals from different species but when that is done the offspring are sterile. For example, a mule is the sterile offspring of crossbreeding a horse with a donkey. According to Dr. Morris

> The small or *micro*evolutionary changes occur by recombining existing genetic material within the group. As Gregor Mendel observed with his breeding studies on peas in the mid 1800's, there are natural limits to genetic change. A population of organisms can vary only so much.

Here is where the failed "upward reasoning" referred to by Dr. Mayr comes in. Evolutionists take those known types of micro changes, postulate beneficial mutations that have never been observed, theorize "punctuated equilibrium" resulting from stored up mutations, mix in a few billion years, and triumphantly declare for our belief that it scientifically explains macroevolution, the supposed origin of all of the unique types of living things. Dr. Morris explains:

> *Macro*evolution refers to major evolutionary changes over time, the origin of new types of organisms from previously existing, but different, ancestral types. Examples of this would be fish descending from an invertebrate animal, or whales descending from a land mammal. The evolutionary concept demands these bizarre changes. Evolutionists assume that the small, horizontal *micro*evolutionary changes (which are observed) lead to large, vertical macroevolutionary changes (which are never observed). This philosophical leap of faith lies at the eve of evolution thinking.

**Mutations Remove Genetic Information**

Genetic mutations do occur because of what might be thought of as "copying errors" in the DNA code. These are not beneficial because they don't add information, they remove information. Once they occur they can be passed along through reproduction. A genetic disorder is a condition caused by an absent or defective gene or by a chromosomal aberration. The NIH's Human

Genome Project has identified specific human gene flaws as markers for certain diseases or conditions. On the other hand, beneficial mutations have not been observed. One that has often been cited in evolution propaganda is sickle-cell anemia that provides an individual with enhanced resistance to malaria. However, sickle-cell anemia is a serious and sometimes fatal blood disorder. See http://creation.com/exposing-evolutions-icon. Dr. Morris explained how textbooks use examples of microevolution to "sell" belief in macroevolution to children:

> A review of any biology textbook will include a discussion of *micro*evolutionary changes. This list will include the variety of beak shape among the finches of the Galapagos Islands, Darwin's favorite example. Always mentioned is the peppered moth in England, a population of moths whose dominant color shifted during the Industrial Revolution, when soot covered the trees. While in each case, observed change was limited to *micro*evolution, the inference is that these minor changes can be extrapolated over many generations to *macro*evolution.

In that paragraph Dr. Morris mentioned the style of evolutionary writing that this writer pointed out earlier in his critique of the PBS book's section "In Search of Origins." Noted was the way words and pictures were used so that a reader would draw a conclusion by inference that the written words did not actually state. This is what is done to school children when microevolutionary examples are provided as the mechanism of macroevolution.

## Darwin's Finches in Fiction Forever

In the paragraphs above, Dr. Morris mentioned that "A review of any biology textbook will include a discussion of *micro*evolutionary changes. This list will include the variety of beak shape among the finches of the Galapagos Islands, Darwin's favorite example."

The telling of the evolution story often begins with how Charles Darwin, as a voyager on HMS Beagle, stopped in the Galapagos Islands and collected finches, and by observation of different beak sizes, he developed his theory. Darwin collected finches and other birds on that trip but he didn't know anything much about birds. When he got back to England in 1837, he turned the preserved birds over to a bird expert, John Gould, who proclaimed that the finches represented 12 distinct species. That was reported in newspapers of the day. Subsequently, "Darwin's Finches" became icons of evolution, although they are simply examples of horizontal variation or adaptation. The finches have been considered textbook examples of how a single species turned into many species to exploit different resources. Subtle changes in size and structure of beaks among the species of ground finches have been called "evolution caught in the act." Although this is actually nothing more than natural selection operating on an existing, information-rich gene pool, that entailed the rearrangement and/or loss of existing genetic information from populations, Darwinists like to call it evolution in school textbooks, using microevolution to "sell" macroevolution to children.

In an online "course for educators," PBS teaches that

> Darwin thought that evolution took place over hundreds or thousands of years and was impossible to witness in a human lifetime. Peter and Rosemary Grant have seen evolution happen over the course of just two years.

http://www.pbs.org/wgbh/evolution/educators/course/session4/elaborate_b_pop1.html

Princeton University ornithologists (bird experts), Peter and Rosemary Grant, have been spending half a year in the Galapagos, observing finches for decades. Based on their work, in 1994, Jonathan Weiner published *The Beak of the Finch: A Story of Evolution in Our Time.* According to a reviewer of that book,

> Darwin's finches exhibit an unusually high degree of variability. This, coupled with the fact that the Grants and their co-workers were fortunate enough during their 20-year vigil to experience a severe drought and the very opposite, means that it is no surprise that they were able to document some quite rapid changes under [natural] selection. When the drought brought a shortage of easily available small seeds, is it any wonder that the birds with big beaks survived better because they were the only ones to be able to crack big seeds, and so on?
>
> After all the 'hype' about watching 'evolution', one reads with amazement that the selection events observed actually turned out to have no net long-term effect. For example, for a while [food-dependant] selection drove the finch populations towards larger birds, and then when the environment changed, it headed them in the opposite direction. The author says concerning this sort of effect (also seen in sparrows) that 'Summed over years, the effects of natural selection were invisible' (p. 108). So that when Darwin looked at the fossil record and found it 'static and frozen for long stretches' (p. 109), this was the reason.
>
> Evolutionists have long argued the opposite—that evolution is invisible in the short term, but would become visible if we had enough time. Yet according to Weiner, we can see evolution happening in the (very) short term, but any longer and it becomes 'invisible'! The mind

boggles at how evolutionists can be blind to this inconsistency.http://creation.com/book-review-the-beak-of-the-finch

Recommended reading, "Darwin's Finches: The hype continues." evolutionnews.org/2016/04/post_44102796.html

The PBS propaganda "course for educators" is based on the same fallacy promoted by Weiner's *Beak of the Finch.* It does not make the distinction between microevolution and macroevolution. It labels as evolution natural selection operating on an existing, information-rich gene pool. It is the rearrangement and/or loss of existing genetic information from populations. The result of not making the distinction between micro and macroevolution is to mislead. What is needed is a "Truth in labeling" campaign applied to education.

**Defending the Icon**

As reported in the April 2015 edition of *Discover*, 21st Century science utilizing DNA indicates that none of the "species" are distinct. Robert Zink of the U. of Minnesota's Museum of Natural History is an ornithologist who said that sequences of their nuclear and mitochondrial DNA show little variation, and none of the telltale signs that suggest distinct species. Zink said it makes more sense to classify the birds as a single species of ground finch with ecologically-driven variation. (The changes are "horizontal" within the species; there is no "vertical" descent.)

Faced with the results of modern genetic analysis, showing that the birds are one species, according to the *Discover* article, the Grants replied that the birds are "on their way to becoming separate species." In other words, they are in the "evolutionary stasis" phase of "punctuated equilibrium," but in a million or so years a "Hopeful Monster" will pop up.

## That Microevolution Does Not Explain Macroevolution Is Old News to Evolutionary Biologists

It is a fundamental observation of humans that effects have causes. What are the supposed causes of macroevolution? In November 1980, a conference of some of the world's leading evolutionary biologists, billed as 'historic,' was held at the Chicago Field Museum of Natural History on the topic of 'macroevolution.' Reporting on the conference in the journal *Science* (Vol. 210 (4472):883–887, 1980.), Roger Lewin wrote: "The central question of the Chicago conference was whether the mechanisms underlying microevolution can be extrapolated to explain the phenomena of macroevolution. At the risk of doing violence to the positions of some of the people at the meeting, the answer can be given as a clear, No."

Francisco Ayala (Associate Professor of Genetics, University of California), was quoted in the same article as saying "… but I am now convinced from what the paleontologists say that small changes do not accumulate." However, Ayala, the failed Dominican priest, continued as a life-long promoter of materialistic, random chance evolution.

Examples of microevolution are still being used in textbooks to fool children into believing what all evolutionary biologists know is false. A woman who wrote to the Institute for Science and Catholicism from Minnesota in January 2016 said:

> Common Core has hit us hard here. Tenth grade biology includes a 6-week program on evolution beginning with the teacher exclaiming that "I'm sorry, but if any of you or your parents think creation has anything to do with it, you're wrong. There is a scientific consensus that the evidence is irrefutable."

Is there anybody reading this ready to help save Catholic children from these faith-destroying lies? A priest, a Catholic intellectual, anybody?

**Evolution as Religion**

From the numerous examples given above one can see that all evolutionary speculations qualify as "science" (as long as they exclude God). Yet evolutionism fills the gap of "no God" because it answers for those who believe it "questions of ultimate concern" that are beyond science and formerly were considered religion and philosophy. C.S. Lewis was not the first to notice this but he often remarked about it. For example, in a 1944 address to the Oxford University Socratic Club he said:

> More disquieting still is Professor D. M. S. Watson's defense. "Evolution itself," he wrote, "is accepted by zoologists not because it has been observed to occur or... can be proved by logically coherent evidence to be true, but because the only alternative, special creation, is clearly incredible." Has it come to that? Does the whole vast structure of modern naturalism depend not on positive evidence but simply on an a priori metaphysical prejudice? Was it devised not to get in facts but to keep out God?

British philosopher Mary Midgley recognized long ago that

> Evolution is the creation myth of our age. By telling us our origins it shapes our views of what we are. It influences not just our thoughts but also our actions in a way which goes far beyond its official function as a biological theory.

In the next chapter readers will learn what the Magisterium teaches about evolution. In Chapters 9 and 13 readers will see that Cardinal Ratzinger identified evolution as a philosophy contrary to Catholicism.

# Chapter 8-*Humani Generis* Explained

Those warnings about Fr. de Chardin's books came years after it was obvious that, to the detriment of themselves and their students, the Marist Brothers, Fordham Jesuits, and teachers everywhere were also ignoring since 1950 the binding teaching of Pope Pius XII's encyclical *Humani Generis.* The English title of that encyclical is *The Human Race: Some False Opinions Which Threaten To Undermine Catholic Doctrine.* It is quite possible that the Pope wrote this encyclical at that point in time, 1950, because he was aware that works such as those produced by Fr. de Chardin and others were being published or circulating both "above and under the table" in Catholic institutions.

Many sincere Catholics who perhaps have never read that encyclical entirely, or who have been misled about its content by Catholic teachers too smug or lazy to read it, believe that it said that Pope Pius XII in paragraph 36 of the encyclical permitted belief in evolution. For example, the perfectly orthodox Catholic philosopher, John Young, a lecturer at the Center for Thomistic Studies in Australia, asserted as much in a January 29, 2015, debate via newspaper with Hugh Owen and other scholars from the Kolbe Center for the Study of Creation. When Young was asked in a letter to the editor of that same newspaper if he meant by "evolution" what the scientific consensus means by evolution or if he meant non-controversial microevolution such as selective breeding of dogs, he did not answer directly. Mr. Young said that he held that materialistic macroevolution is totally impossible. Theistic evolutionists do not believe in purely materialistic evolution because they believe God guided it but they provide no details, scientific or theological. Young's opinion that Pope Pius XII permitted belief in evolution, with or without God's guidance, remains unjustified. In this chapter, the Pope's own words will disprove that notion. Fr. Warkulwiz, in his book *Humani Generis on Evolution: Reading It Completely and Consistently*, observed that

> Many Catholics seem to think that all the church has ever said pertaining to evolution is contained in paragraph 36 of *Humani Generis*...they misinterpret that passage and make it the Magna Carta giving Catholics liberty to profess and promote belief in biological evolution, which it certainly is not.

*Humani Generis* is primarily about the increasingly bad philosophy spreading through Catholic institutions of formation because of the uncritical acceptance of evolutionary theory and the need for Catholic theologians and philosophers to combat it by rediscovering and teaching the philosophy of Doctors of the Church such as St. Thomas Aquinas. According to Cardinal Ratzinger's assessment decades later, the situation has gone from bad to worse as the reader will see. Pius XII identified evolution as the problem in the 5th paragraph of the encyclical. Consider these opening paragraphs:

> 5. If anyone examines the state of affairs outside the Christian fold, he will easily discover the principle trends that not a few learned men are following. Some imprudently and indiscreetly hold that evolution, which has not been fully proved even in the domain of natural sciences, explains the origin of all things, and audaciously support the monistic and pantheistic opinion that the world is in continual evolution. Communists gladly subscribe to this opinion so that, when the souls of men have been deprived of every idea of a personal God, they may the more efficaciously defend and propagate their dialectical materialism.

> 6. Such fictitious tenets of evolution which repudiate all that is absolute, firm and immutable, have paved the way for the new erroneous philosophy which, rivaling idealism, immanentism and pragmatism, has assumed the name of existentialism, since it concerns itself only with existence of individual things and neglects all consideration of their immutable essences.

It does not appear to this writer that the Pope was writing to permit belief in evolution. Some younger persons not familiar with Communism and the antithesis of the Church toward it might miss the significance of the Pope's reference to "Communists" in paragraph 5 above. In 1937 his predecessor, Pius XI, wrote an encyclical, *Divini Redemptoris* (On Atheistic Communism), reinforcing the long-standing Church teaching against that ideology and its connection to evolution.

> The doctrine of modern Communism, which is often concealed under the most seductive trappings, is in substance based on the principles of dialectical and historical materialism previously advocated by Marx, of which the theoricians of bolshevism claim to possess the only genuine interpretation. According to this doctrine there is in the world only one reality, matter, the blind forces of which evolve into plant, animal and man. Even human society is nothing but a phenomenon and form of matter, evolving in the same way. ... In a word, the Communists claim to inaugurate a new era and a new civilization which is the result of blind evolutionary forces culminating in humanity without God.

In 1950, when Pius XII wrote *Humani Generis*, armies of the NATO Alliance and the Communist Soviet Union's Warsaw Pact armies stood eyeball to eyeball at a border of thousands of miles, known as "The Iron Curtain," which ran through Europe and enclosed the eastern part of that continent under Communist rule. It is hard to imagine a more serious warning about evolution could have been made in 1950 than to explain how useful it was to the Communists. Pope Pius knew where evolutionary theory leads when it undergirds a philosophy. He said that through it "the souls of men have been deprived of every idea of a personal God."

### What the Pope Wrote and Why

Those negligent teachers ("blind guides") and misled students who maintain that *Humani Generis* permitted Catholics to believe in evolution claim that permission is granted in paragraph 36. Below is paragraph 36 in its entirety but put into context by paragraphs 34 and 35.

> 34. It is not surprising that these new opinions endanger the two philosophical sciences which by their very nature are closely connected with the doctrine of faith, that is, theodicy and ethics; they hold that the function of these two sciences is not to prove with certitude anything about God or any other transcendental being, but rather to show that the truths which faith teaches about a personal God and about His precepts, are perfectly consistent with the necessities of life and are therefore to be accepted by all, in order to avoid despair and to attain eternal salvation. All these opinions and affirmations are openly contrary to the documents of Our Predecessors Leo XIII and Pius X, and cannot be reconciled with the decrees of the Vatican Council. It would indeed be unnecessary to deplore these aberrations from the truth, if all, even in the field of philosophy, directed their attention with the proper reverence to the Teaching Authority of the Church, which by divine institution has the mission not only to guard and interpret the deposit of divinely revealed truth, but also to keep watch over the philosophical sciences themselves, in

order that Catholic dogmas may suffer no harm because of erroneous opinions.

35. It remains for Us now to speak about those questions which, although they pertain to the positive sciences, are nevertheless more or less connected with the truths of the Christian faith. In fact, not a few insistently demand that the Catholic religion take these sciences into account as much as possible. This certainly would be praiseworthy in the case of clearly proved facts; but caution must be used when there is rather question of hypotheses, having some sort of scientific foundation, in which the doctrine contained in Sacred Scripture or in Tradition is involved. If such conjectural opinions are directly or indirectly opposed to the doctrine revealed by God, then the demand that they be recognized can in no way be admitted.

36. For these reasons the Teaching Authority of the Church does not forbid that, in conformity with the present state of human sciences and sacred theology, research and discussions, on the part of men experienced in both fields, take place with regard to the doctrine of evolution, in as far as it inquires into the origin of the human body as coming from pre-existent and living matter - for the Catholic faith obliges us to hold that souls are immediately created by God. However, this must be done in such a way that the reasons for both opinions, that is, those favorable and those unfavorable to evolution, be weighed and judged with the necessary seriousness, moderation and measure, and provided that all are prepared to submit to the judgment of the Church, to whom Christ has given the mission of interpreting authentically the Sacred Scriptures and of defending the dogmas of faith. Some however, rashly transgress this

liberty of discussion, when they act as if the origin of the human body from pre-existing and living matter were already completely certain and proved by the facts which have been discovered up to now and by reasoning on those facts, and as if there were nothing in the sources of divine revelation which demands the greatest moderation and caution in this question.

## It Must Be Read in Context

In context, one can see that the Pope is pointing out in 34 that the "new opinions" are contrary to the constant teaching of the Church and to illustrate his point he names Leo XIII, Pius X, and the First Vatican Council. In paragraph 35, he points out how questions of the positive sciences are connected with the Faith and that some in the Church insistently demand that our religion take science into account. He said that those demands would be praiseworthy if the science involved clearly proved facts. But, he says, when the demands concern mere hypotheses having some sort of scientific foundation and are conjectural opinions opposing revelation from God, the demand can in no way be admitted. So in paragraph 36, for the reasons stated in 35, namely, that voices within the Church are demanding action on the part of the Church to accept unproved assertions of facts and hypotheses just because they have some sort of scientific foundation, he set up conditions under which the claims of science can be evaluated through research and discussion. In paragraph 36 he set rules for such research and discussions.

1. All those involved must be experienced in both the present state of human sciences and sacred theology.
2. Any research and discussion by those experts in both fields that proposes evolution was involved in the origin of the human body must begin with the principle that, as a minimum, it began with pre-existing and living matter and that souls are immediately created by God.

3. With respect to whatever hypothesis or assertion of fact regarding evolution that was being considered, favorable and unfavorable opinions must be heard. Those expressing opinions must support them with reasons that can be weighed and judged with seriousness, moderation and measure.
4. As a qualifying condition to participate in any research and discussions, participants must acknowledge that the Church, not they, will make the authoritative judgments and that they must be willing to assent to whatever is decided.

Nowhere in those conditions read in the context of paragraphs 34 and 35 can one read what Jesuits taught Mr. Fitzpatrick and lay intellectuals who pride themselves on their orthodoxy are still teaching.

### Did Theologians, Philosophers, and Other Teachers Adhere to Those Conditions?

With respect to condition #1 above, it has been mostly theologians and philosophers with little or no scientific expertise telling other Catholics what they can believe about evolution based on the bogus "science" they were taught in high school. There are college theology and philosophy faculty members who have been teaching from the same set of notes for 30 years and one wonders if they have read anything about the discoveries natural science has made since they were in high school.

Condition #2 is the "show stopper," so to speak. Anyone who thinks that this condition intended to give permission to believe in evolution of humans should go back and read what Pius said about evolution in paragraphs 5 and 6. Some fiat creationists say #2 was a "compromise" that went against constant Church teaching insofar as it did not forbid research and discussion of the possibility that Adam's body and soul were not brought into existence at the same time. For example, St. Thomas Aquinas in his *Summa Theologia* taught that the body and soul of Adam were brought into being at the same time. I think Pius found a brilliant way of challenging the Catholic evolutionists who he said (in paragraph 35) "insistently demand that the Catholic religion take these sciences into account as much as possible;" he challenged them to "put up or shut up." Pius XII knew his infallibility did not extend to science so he did not make a scientific statement in paragraph 36. Instead he made a binding theological statement regarding the infusion of a soul that is absolutely unacceptable to the scientific consensus and could never be reconciled with it

The scientific consensus rejects the spiritual soul infused by God and when the evolutionist propositions are accepted as true science, the materialist explanation of human uniqueness gets pretty bizarre. For example, consider the teaching of Stephen Jay Gould as it relates to that mystery. Gould's opinions must surely represent the scientific consensus in 2002 because of who he was. In his 2002 book, Professor Gould told how the evolving thing made that step forward to that immaterial something generally described as human consciousness. Referring to the "vertebrate brain," he wrote that "evolution grafted consciousness in human form upon this organ in a single species." (*I Have Landed: The End of a Beginning in Natural History*, p. 55.) So, according to a professor who was most surely in the mainstream of evolutionary science, a process called evolution which cannot be proved or disapproved, but is believed to have happened by the scientific consensus, by some means or other, had at its disposal "consciousness in human form" which the process then grafted on to the brain of some unspecified vertebrate. If it were to be said that "God grafted consciousness in human form upon this organ in a single species," Gould and his science colleagues would say that is "just religion," not science. Your brain is "Beyond Belief." Read https://www.icr.org/article/10186

### What Happened in Response to Paragraph 36?

With respect to both condition #2 and #3 of "not forbidden" discussions, Pius XII pointed out in paragraph 36 that the conditions were going to be challenging for Catholic evolutionists to live with. He pointed out that there are those who "rashly transgress this liberty of discussion, when they act as if the origin of the human body from pre-existing and living matter were already completely certain and proved by the facts which have been discovered up to now and by reasoning on those facts, and as if there were nothing in the sources of divine revelation which demands the greatest moderation and caution in this question."

Catholic evolutionists did not "put up" and have not "shut up." It has been more than 65 years since research and discussion under the conditions specified by the Magisterium could have begun, but as far as the supporters of the scientific consensus and its Humanist philosophy are concerned, there was nothing to discuss. The aggressor side in any war is usually not motivated to offer terms, especially if he has infiltrated the opponent's ranks as deeply as Humanists have infiltrated Catholic departments of philosophy, theology, and Bible studies. In technical papers of the 1940s, the modern Darwinian orthodoxy had not yet congealed and a style of doubt remained quite common among evolutionary biologists, according to Stephen Jay Gould. In an article he wrote in 1983 called, "The Hardening of the Modern Synthesis," he documented that evolution only coalesced as the hard-line orthodoxy in the 1950s and 1960s.

The scientific consensus moved into nearly all of the public schools in that timeframe, in part, because the Federal

Government's Biological Sciences Curriculum Study strongly recommended that evolution be made the central focus of the study of biology at all levels. New biology textbooks were written reflecting this view and they were quickly adopted by many states. Already in the 1960s, evolution became a "must believe" tenet of the Government's ruling, yet uncrowned, State Religion, namely, Humanism. According to the PBS Evolution program companion book, *evolution The Triumph of an Idea*, it was the launch of the first earth satellite, Sputnik, by the Soviet Union in 1957 that gave the opportunity to the Humanist Federal bureaucrats to impose evolution on the children in the States that had resisted.

> The triumph of Soviet science created a national panic over the state of American science education—including the teaching of evolution. Textbooks began surveying evolution again, and by 1967 even the Tennessee legislature had repealed the law that had gotten Scopes arrested.

Picture Werner Von Braun, father of America's rocket program, sitting at his desk at the Marshall Space Flight Center in Huntsville, Alabama. He is wringing his hands over the Soviet achievement and exclaiming "what I really need are some more evolutionary biologists to help me get our Saturn rocket powerful enough to put a satellite into orbit." Coincidentally, this writer was getting a degree in electrical engineering between 1958 and 1962. Evolution was never mentioned. Upon graduation, his first offer of employment was with one of Von Braun's engineering contractors in Huntsville and they never asked what he knew about evolution.

While there has been no movement toward reconciliation with Catholic teaching by the scientific consensus, the teaching organs of Catholic institutions have either deferred to the opinions of "science" or maintained a respectful silence, according to Cardinal Joseph Ratzinger in a speech to be quoted in some detail later. Leading Catholic philosophers, theologians, and teaching authorities left the field of battle or defected to the other side.

## Adam was Not a Population

In paragraph 37, Pope Pius XII made another binding theological statement that is incompatible with the general theory of evolution, namely, that all humans are descendent from one man, Adam.

> For the faithful cannot embrace that opinion which maintains that either after Adam there existed on this earth true men who did not take their origin through natural generation from him as from the first parent of all, or that Adam represents a certain number of first parents.

That is called monogenism. The opinion that humans are descended from more than one man is called, by the Pope, polygenism. He called polygenism a "conjectural opinion" that the "children of the Church" may not hold. That teaching could never be accepted by the scientific consensus. According to the general theory of evolution, there is no reason why more than one evolving thing should not have evolved into a human about the same time. As Fr. Warkulwiz observed about paragraph 37 in his book on *Humani Generis*:

> Here Pius XII further undermines further advocacy of human evolution by Catholics. Evolutionists generally think of evolution as taking place in populations, not individuals. They consider evolution through an individual as unlikely.

In other words, evolutionists do not hold that only one unique evolving thing or only one unique evolving thing couple made an "evolutionary breakthrough."

## What Evolution Isn't

Catholics who doubt that the scientific consensus believes exactly what Fr. Warkulwiz said it believes are referred to *What Evolution Is* by the "Darwin of the 20th Century," Dr. Ernst Mayr. In a chapter called "How and Why Does Evolution Take Place?" he asks:

> What evolves? Do individuals evolve? Certainly not in any genetic sense…Then what is the lowest level of living organization to evolve? It is the population. And the population turns out to be the most important site for evolution. Evolution is best understood as the genetic turnover of the individuals of every population from generation to generation.

In 1859 when Darwin's theory was published, little was known about genetics, so it was accepted that it was individuals who had developed variations which they passed on to the next generation who were the engines of evolution. Dr. Mayr knew there were no transitional fossils. He was at Harvard with Stephen Jay Gould. What Mayr is alluding to here is Gould's punctuated equilibrium which was supposed to take place in populations.

Dr. Mayr was quoted to support Fr. Warkulwiz's statement that evolutionists consider evolution through one individual as unlikely. This means polygenism. That Catholics may not believe in polygenism is a formal teaching of the Catholic Church.

## Polygenism is a Humanist Dogma

Dr. Mayr, Stephen Jay Gould and other Humanist scientists hold with Darwin that man descended from animals. That means there was biological continuity from animals to humans. If *Genesis* is read according to the ordinary meaning of words, then the formation of Adam from the dust of the ground speaks of a direct and unique act by God with a definite discontinuity between

Adam and the other animals. Evolutionist bible scholars have reinterpreted *Genesis* to comply with the opinion of Humanist science. (The harm done by such scholars will be put into perspective by a speech made by Cardinal Ratzinger and quoted in the next chapter.)

A representative example of reinterpretation of the Bible by evolutionist scholars is given in the work of J.H. Walton. Walton is a prominent author who in 2001 became Professor of Old Testament at Wheaton College, a liberal Protestant facility. In his 2009 book *The Lost World of Genesis One* and in a published piece, "An Historical Adam: Archetypal Creation View", he perfectly illustrated the Humanist bible scholars' adaptation to the claims of evolutionary science. Walton believes that mention of the formation of Adam from the dust of the ground in *Genesis* 2 should be read as archetypal and not as prototypal of humanity. That is, Adam is seen as just a representative of other people alive at the time, and the dust refers to Adam's mortality, and not to material substance. He thinks the passage is therefore using functional language about human frailty and not speaking materially about a literal creation. This position holds that Adam may have had a human mother and father and there is no material discontinuity with animals as suggested by a literal reading of the text.

The evolutionary interpretation does violence to Catholic teaching regarding Original Sin. Paragraph 416 of the *Catechism of the Catholic Church* teaches that "By his sin, Adam, as the first man, lost the original holiness and justice he had received from God, not only for himself but for all human beings." That teaching is what Pius XII was conserving when he ruled out two contrary opinions by *Humani Generis* paragraph 37:

> For the faithful cannot embrace that opinion which maintains that either after Adam there existed on this

> earth true men who did not take their origin through natural generation from him as from the first parent of all, or that Adam represents a certain number of first parents.

Does Walton's view that there were other people alive who did not take their origin from Adam and that Adam may have had a mother and father negate the Immaculate Conception as unique?

**Evolving Things and the Immaculate Conception**

Some Catholics overlay the 19th Century "science" with God's intervention to arrive at conclusions such as "Catholics are free to believe that evolution took place, as long as they understood it to be a process begun by God, and one in which human beings were created when God infused a soul into the evolving creature that became man." A good example of that crossing of bogus science with bad theology is an article published March 28, 2017 in the *National Catholic Register* called "How Do Adam and Eve fit with Evolution?" The author, a chemistry Ph. D. who writes "Catholic-science" books and has a blog wrote that

> What are we sure of? We can say that God created our first parents, as He did all creatures, and that they were highly complex organisms. That description applies whether Adam and Eve began as zygotes with human souls growing in maternal bodies or as naked adults in a garden.

If Adam and Eve or those evolving things were zygotes with human souls growing in maternal bodies, wouldn't that make them the first Immaculate Conceptions because they were certainly conceived without Original Sin? And if Adam had a mother, how can we square that with Luke's gospel in which he gives the genealogy of Jesus by naming the fathers all the way back to Adam who he says was "the son of God."? (3: 38)

## Adam's Rib

As shall be explained in Chapter 14, modern Humanism was conceived in the 19th Century and came to full maturity early in the 20th Century in the universities, if not in the street. It is not that the Church ignored it. The Church's responses began with the Decree from the Provincial Council of Cologne in 1860, the encyclical of Leo XIII in 1893, and Decisions of the Pontifical Biblical Commission (PBC) of 1909. For an excellent historical and theological account of the Church's response only made possible in recent decades when the Congregation for the Doctrine of the Faith made available its archives up to the end of 1903 see "Early Vatican Responses to Evolutionist Theology" http://www.rtforum.org/lt/lt93.html

Among the decisions of the PBC is the statement that Catholics may not question the formation of the first woman from the first man. That was reaffirmed by paragraph 371 of the *Catechism of the Catholic Church* in 1994. That teaching creates additional problems for Catholics who combine the evolutionary scientific consensus with the infusion of a soul into one or more evolving things. If Adam was created by the infusion of a soul into an evolving thing, how did the first woman come to be? If one suggests that the first woman also resulted from the infusion of a soul into a different type of evolving thing that would negate "Adam as the first parent of all," and make her also an Immaculate Conception. If one suggests that the first woman's body did not evolve but was specially made by God as the Bible says, wouldn't that confer higher dignity on the first woman than on the first man? That can't be reconciled with the teaching of complementary but equal dignity of men and women as affirmed by paragraph 369 of the *Catechism of the Catholic Church.* 1Timothy 2:13 tells us "Adam was formed first, then Eve."

Why did God use Adam's rib? There is a possible answer in an article called "Regenerating ribs: Adam and that 'missing' rib."

Read it at http://creation.com/regenerating-ribs-adam-and-that-missing-rib .

## Teaching Evolution Prohibited

The most serious question regarding the sincerity or fidelity of those Catholics who, for whatever reason, taught that paragraph 36 authorized Catholics to believe in evolution is provoked by paragraphs 41 and 42 of that encyclical. They were told not to teach evolution and this command was a morally binding directive.

> 40. …But we know also that such new opinions can entice the incautious; and therefore we prefer to withstand the very beginnings rather than to administer the medicine after the disease has grown inveterate.
>
> 41. For this reason, after mature reflection and consideration before God, that We may not be wanting in Our sacred duty, We charge the Bishops and the Superiors General of Religious Orders, binding them most seriously in conscience, to take most diligent care that such opinions be not advanced in schools, in conferences or in writings of any kind, and that they be not taught in any manner whatsoever to the clergy or the faithful.
>
> 42. Let the teachers in ecclesiastical institutions be aware that they cannot with tranquil conscience exercise the office of teaching entrusted to them, unless in the instruction of their students they religiously accept and exactly observe the norms which We have ordained. That due reverend and submission which in their unceasing labor they must profess toward the Teaching Authority of the Church, let them instill also into the minds and hearts of their students.

## Man-Made Doctrine Taught

We now know all too well how paragraphs 41 and 42 were obeyed, and the disastrous consequences of their disobedience, particularly the overt disobedience of some priests in religious orders. But it must also be noted that there are Catholic institutions of higher learning that pride themselves as being orthodox and yet teach that it is acceptable to believe in evolution overlaid with God's direction. One institution that takes pride in its fidelity, and at which the entire lay faculty pledges fidelity to the Magisterium on an annual basis, has been teaching an interpretation of Church doctrine which, when combined with the students' incoming high-school education in support of the scientific consensus, leads many students to become non-scientific, non-biblical hybrid theistic evolutionists who think it "works" for them. Taught is a hodgepodge of Catholic doctrine and opinions of theologians and philosophers mixed with the conventional wisdom of the "evolution story." It is not "scientific," for no scientist would hold this theory, and it is not biblical, because it discounts the truth of the inerrancy of Scripture. In short, what is being taught is not anything the Church has ever held to be true.

In March 2017, for example, the Theology Chair at this particular college invited speakers from the Kolbe Center for the Study of Creation to give an evening presentation. The presentation was to supplement the coursework from a class that he was teaching called "Genesis, Creation, and Evolution." The Kolbe Center promotes *fiat* creationism, and backs it up with highly-qualified natural science speakers, so the professor thought that the students would benefit from hearing this side of the creation/evolution debate. When word got out that this presentation was going to take place, the theology professor was pressured by members of the faculty and administration to cancel the invitation, but he "stuck to his guns." He was allowed to invite his own students to attend, but no campus-wide publicity was permitted. Subsequently, just a few weeks later, a philosophy professor was given campus-wide publicity for his "special talk on evolution" in which, according to the promotional material, he offered "his thoughts on why evolution is compatible with theism (with a belief in God)." This "faithful" Catholic college needs a reality check on its promotion of itself as orthodox.

Thaddeus Kozinski, Academic Dean and Associate Professor of Philosophy at Wyoming Catholic College took theistic evolution-teaching academics to task in an essay called "Catholic Education and the Cult of Theistic Evolution." He noted that

> I am speaking of the Catholic theistic evolutionists. They overstep *science's* bounds when they claim that debatable theories, such as the theory of evolution, are "facts"—something that Pius XII condemned very unequivocally with regard to evolution in *Humani Generis*. They overstep science's bounds again when they attempt to render certain non-verified, non-facts, such as common descent from mono-celled organisms, as verified, indisputable facts by recourse to, not actual indisputable evidence, but the social force of the so-called "scientific consensus," that same force that fires and character-assassinates people who publish peer-reviewed scientific articles that conclude to, say, intelligent design of certain cellular processes, and that excludes anyone but committed evolutionists to the Pontifical Academy of Sciences. They overstep *philosophy's* bounds when they teach debatable and idiosyncratic philosophical theories about causality in the natural world and its relation to God, claiming, for example, that God's providence over the world is compatible with genuine chance in nature—yes, not just the *appearance* of chance, but chance!—as if this were the only rational and Thomistic way to explain things, as if serious and sophisticated philosophical challenges to it, such as found in the work of Robert

Koons, are just, a priori, otiose and tending towards fundamentalism. They overstep *theology's* bounds when they dismiss the very serious challenges, not just to evolutionary theory, but to the very fact of evolution itself, from not only the Catholic Magisterium and Fathers of the Church, but also from the latest scientific evidence, which has, it must be said, proved neither common descent of humans from primitive organisms, nor the generation of all life, in all of its glorious complexity and design, from mindless natural selection conserving random genetic variation and mutation.

http://www.theimaginativeconservative.org/2015/06/catholic-education-and-the-cult-of-theistic-evolution.html

**Theistic Evolutionists and Thomas Aquinas**

Many theistic evolutionists cling to their belief by asserting that Thomas Aquinas supported the idea that evolution can be reconciled with Faith. Michael Chaberek, O.P., S.T. D., chaplain at Thomas Aquinas College (TAC), may have buried that excuse with his 2017 book, *Aquinas and Evolution.* David Arias, a philosophy professor at Our Lady of Guadalupe Seminary, has written that Father Chaberek made an excellent textual, philosophical, and theological case that the teaching of Aquinas "on human origins is incompatible with macroscopic evolutionary theory, even so-called 'theistic evolution'."

The Foreword to Chaberek's book was written by a philosophy professor at the Franciscan University (Steubenville). Are those intellectuals at Wyoming Catholic, TAC, Our Lady of Guadalupe Seminary, and Steubenville, the 21st Century evangelism leaders Catholics have so badly needed?

# Chapter 9-Biblical Interpretation

This chapter illuminates the impact that belief in evolution has had on the decrease in belief in the Bible. A lecture was delivered January 27, 1988, at Saint Peter's Church in New York by Cardinal Ratzinger, Prefect of the Congregation for the Doctrine of the Faith, who became Pope Benedict XVI. The title was "Biblical Interpretation in Crisis: On the Question of the Foundations and Approaches of Exegesis Today." It echoed the same type of criticisms of Catholic scholars that Pope Pius XII had made, 38 years earlier, in his encyclical *Humani Generis*, namely, the pollution of Catholic philosophy and theology by belief in evolution. Here is one paragraph from that lecture that pinpoints how evolution perverts Biblical interpretation:

> In the first place, one can note that in the history-of-religions school, the model of evolution was applied to the analysis of biblical texts. This was an effort to bring the methods and models of the natural sciences to bear on the study of history. Bultmann laid hold of this notion in a more general way and thus attributed to the so-called scientific worldview a kind of dogmatic character. Thus, for example, for him the nonhistoricity of the miracle stories was no question whatever anymore. The only thing one needed to do yet was to explain how these miracle stories came about. On one hand the introduction of the scientific worldview was indeterminate and not well thought out. On the other hand, it offered an absolute rule for distinguishing between what could have been and what had to be explained only by development. To this latter category belonged everything which is not met with in common daily experience. There could only have been what now is. For everything else, therefore, historical

processes are invented, whose reconstruction became the particular challenge of exegesis.

### Examples of the Scientific Method

Leo XIII wrote against those "scientific method" folks in 1893. Some examples below will put a face on what Cardinal Ratzinger was talking about when he said "the model of evolution was applied to the analysis of biblical texts." Bultmann, to whom Cardinal Ratzinger referred, was the German Protestant Rudolf Bultmann. He was born in 1884. He became a university professor in 1921, when evolution and non-theistic philosophy were already the rule in German universities. He gained fame as a Biblical exegete. He is credited with the founding of the scientific method also known as the historical-critical method of interpreting the Bible, "to find out how these miracle stories came about." Bultmann's approach relied on his concept of demythology, and he interpreted what he considered "mythological elements" existentially. To catch the flavor of Bultmann's methods, one needs only to consider his explanation of the Resurrection. Bultmann asserted that the resurrection-language of the early Church was used to denote, not a separate event from the crucifixion, but the early disciples' faith that the crucifixion was not a tragic defeat, but the divine act of salvation. Easter is thus about the arising, not of Jesus, but of the faith of the early Church. Jesuits George Tyrrell (1861-1909) and Alfred Loisy (1857-1940), were among the early Catholic scholars who, like Bultman, specialized in the rejection of any idea of supernatural revelation. Loisy, the Biblical Modernist, applied the rejection of divine revelation to Sacred Scripture with Bultmann-like results. Both Tyrrell and Loisy left the priesthood and the Church.

Bultmann's insight regarding that first Easter was taken forward by, for example, the Belgian theologian Edward Schillebeeckx,

O.P., who was born in 1914. According to Fr. Schillebeeckx, when the disciples went to the tomb, their minds were so filled with light that it did not matter if there was a body there or not. What happened in the Easter appearances was a conversion to Jesus the Christ, who now came to them as the light of the world, and this was the "illumination" by which the disciples were "justified."

Pope Francis is also a victim of the scientific method which was part of his evolution-laced Jesuit formation. For example, in July 2015 he applied the Rudolf Bultman technique to the gospel account of the multiplication of the loaves and fishes. According to Jesuit Francis, Jesus didn't miraculously multiply anything. In a homily preached in Christ the Redeemer Square in Bolivia Pope Francis said, "This is how the miracle takes place. It is not magic or sorcery. … Jesus managed to generate a current among his followers: they all went on *sharing what was their own; turning it into a gift for the others*; and that is how they all got to eat their fill. Incredibly, food was left over: they collected it in seven baskets." This writer heard that explanation from a pulpit in Ireland 30 years ago and that helps explain the decline and fall of Catholicism in Ireland. Years ago Francis was sent to a Jesuit facility in Ireland to learn English and maybe that's where he heard that nonsense. One wonders if Pope Francis believes that other "miracle story", namely, transubstantiation. Perhaps he has given what has been interpreted by some national Bishops Conferences to be the "green light" to Communion for those living in adultery because he doesn't believe in the Real Presence. Belief in evolution has serious consequences.

The original and continuing prestige of the historical-critical method depends on its claim to scientific objectivity. Catholics who dispute its findings are immediately charged with ignorance, superstition, fundamentalism, and obscurantism.

### American Practitioner of the Scientific Method

Cardinal Ratzinger identified the problem but did not name names. It is a universal problem. A lay theologian of my acquaintance became a theistic evolutionist upon studying *A Path Through Genesis*, a popular 1956 book by F. Bruce Vawter, C.M. (popularly known as the Vincentian Fathers). Fr. Vawter was ordained in 1946. What follows is to illustrate the way Bible scholars typically helped to propagate belief in evolution to the Catholic population in the years following *Humani Generis,* in spite of being warned against it.

Fr. Vawter was a witness for the plaintiffs who were suing the Arkansas Board of Education in 1981 because they objected to a law passed by the Arkansas Legislature which required the teaching of Creation Science along with Evolution Science in the public schools of Arkansas. From a deposition conducted by the plaintiff's law firm, one learns that Fr. Vawter was taught evolutionary biology and the evolutionary explanation of origins at the St. Thomas Seminary in Denver. When asked if there was any other approach to origins discussed in the classroom besides the evolution approach, he said that there was no conflict in the minds of the people there in "thinking about evolutionary background to the origin of this all and religion." He elaborated:

> I don't think there was any feeling on the part of anybody that there was any incompatibility in presenting it in an evolutionary structure, and at the same time, conceding that the whole thing is not by random decision, but it was a guided or a designed thing, and, therefore, it would not be a question of another model, but rather, evolution would be considered more of the process by which this came to be which would not conflict with the fact it came to be at the behest of a creator.

That non-scientific, non-Biblical combination of evolutionism and theism at Denver in the 1940s is similar to that which Mr. Fitzpatrick said he was taught by other Catholic religious order teachers in the Bronx in the 1950s and 1960s. Fr. Vawter, at the time he gave this deposition, was a Professor and the Chairman of the Department of Religious Studies at DePaul University. His specialty was Old Testament.

**They Were Fr. Vawter's Guides**

In his deposition Fr. Vawter named numerous academic associations to which he belonged and said that their common denominator was the scientific study of religion just by utilizing the scientific method of biblical exegesis (explanation or interpretation of a Bible text). In other words, he was of the Rudolf Bultman school. When asked to name the particular authorities he relied upon in discussing *Genesis* I and II, he replied "all of my predecessors and all the commentators and the accumulated wisdom... that's been amassed in the last couple hundred years in the scientific study of the scriptures." Fr. Vawter spent most of the years of his priesthood until his death in 1986 teaching evolution-polluted Biblical interpretation at various colleges and seminaries. It is useful for Catholic laymen to learn about Fr. Vawter and his history. It gives a glimpse into the academic world of professional, scholar-priests that is typical of the one in which many of our priests and bishops have been educated. Fr. Vawter is not singled out because he was unique but, rather, because he was typical, and many of today's priests and bishops have been educated by similar scholars. In July 2017 a pastor in the Diocese of Arlington, Virginia wrote this:

> In the 1980s, I attended a Midwest seminary that was schizophrenic with respect to the Faith. ... Scripture studies were essentially liberal Protestant. ... One of the Scripture professors, Father Otto, was ... a disciple of Rudolf Bultmann, the famous (or, as I prefer, the

infamous) liberal Protestant theologian whose scholarly technique of "demythologizing" Scriptures corrupted generations of students. Since our Scripture studies were essentially divorced from the Catholic faith, it was only natural that we allowed Protestant seminarians to attend classes.
https://www.thecatholicthing.org/2017/07/02/camp-concentration/

## And the Next Generation

As noted, the late Fr. Vawter was just one example of seminary professors of his generation who undermined Catholicism based on their belief in evolution and passed on their belief to their students. An example from the next generation is Fr. Michael D. Guinan, OFM. He was ordained 18 years after Fr. Vawter. According to Now You Know Media, that sells his publications, he was in September 2015 Professor Emeritus at the Franciscan School of Theology in Berkeley, CA., and had taught at three other seminaries or universities. At age 18, he entered a Franciscan seminary and then went to Catholic University (CUA). He left there in 1972 with an S.T.L in theology and a Ph.D. in Semitic Languages. He was at CUA during the heydays of the priest-scholar open revolt against the Magisterium that began in 1967 led by Fr. Charles Curran of CUA's theology department. That revolt's most destructive point came when about 600 priest-scholars from a variety of Catholic universities and institutions signed their names to a formal declaration of dissent to *Humanae Vitae* written by Curran and CUA theologians in 1968. That declaration was published as paid advertising in major secular newspapers and amplified world-wide by the media. The U.S. Bishops did nothing but "cave." *The Coup at Catholic University* (2015, Ignatius Press) should be read by anyone who is wondering about the beginning of the end of Catholic Higher Education in the U.S. and why the Bishops' moral authority in America lay prostrate before the State when *Roe V. Wade* happened in 1973. Fr. Curran continued as a moral theologian teaching and publishing dissent on sexual matters at CUA until 1986 when Cardinal Ratzinger stripped him of his authority to teach as a Catholic theologian. One can only speculate about the effect of the openly heretical culture at CUA on young Fr. Guinan. Like Fr. Vawter, he chose a career as a biblical scholar and seminary professor. Nothing in his curriculum vitae indicates that he had any formal education, work experience, or any other experience that would qualify him as an authority in natural science. So how, one wonders, did he become (in his own mind) one of those persons described in the last chapter who, with qualifications in both natural science and theology, were "not forbidden" to engage in research and discussion regarding evolution? He would have been in high school in the same years as when James Fitzpatrick was being taught by Marists and Jesuits that evolution was a fact, and their textbooks were filled with hairy half-ape, half-human artist conceptions of their ancestors based on the Piltdown Man forgery.

**U.S. Conference of Bishops and Diocesan Newspapers**

The *Catholic Review* is the newspaper of the Archdiocese of Baltimore. It published an article dated August 28, 2011, titled "Catholic Church has an evolving answer on reality of Adam and Eve" that featured the opinions of Fr. Guinan. The article in the Baltimore newspaper was an excerpt of an article distributed by the U.S. Conference of Catholic Bishops' Catholic News Service (CNS) called "Adam and Eve: Truth or Fiction" by James Breig. The CNS article appeared also in other diocesan papers, such as the *Arlington* (VA) *Catholic Herald* and the Diocese of Brooklyn's *The Tablet*. In that CNS article, the Bishops' journalist Breig wrote that

> In simplified form, the fundamentalist view is that Adam was a real person and the first human created by God, while science argues that human beings evolved as a group. Some Christians hold that they must follow scientific findings and adapt their faith's teachings to that information.

The "fundamentalist" in question is Pope Pius XII. "Science" is atheists like Dr. Ernst Mayr and Dr. Stephen Jay Gould whose "evolution by populations" was part of Gould's now recanted "punctuated equilibrium" theory that replaced the Darwinian theory of "evolution by individuals." Neither theory has stood under the examination of modern scientists currently publishing in peer-reviewed science journals. The CNS and diocesan newspaper editors who printed the story and operate on what they learned in high school can say that both sides were given equal weight. That's what they call "Catholic journalism." The CNS article extensively quoted Fr. Guinan opinion on that subject.

> "Since the 1600s," he said, "the traditional views of Genesis have suffered three challenges: Galileo on the movement of the earth around the sun and not vice versa; the growth of geology in the 18-19th centuries and discoveries about the age of the earth; and Darwin's theory of evolution.

That is the scientific method Bible scholars' typical narrative. At the end of chapter 13 and in Appendix I there is an explanation of how evolutionists use their false version of the Galileo incident to intimidate Catholics, and how Galileo was nothing like the martyr for science that legend has made him. In chapter 6, I explained how 18$^{th}$ and 19$^{th}$ Century geology was made up by Hutton and Lyle. In chapters 4 and 5, I showed that the origin of life and speciation theories of 19$^{th}$ Century Darwinists have never been more disputed than in the 21$^{st}$ Century peer-reviewed research. But does Fr. Guinan know that? He's probably still trapped in the myths he learned in high school and the seminary over 50 years ago about the Piltdown Man, vestigial organs, and whales that lost their legs.

> The church has negotiated these challenges, but not without struggles. Today, no reasonable person in or out of the church doubts any of these three, Father Guinan said in an interview with Catholic News Service.

That is another example of evolutionary propaganda: the classic *ad hominem* fallacy. "All reasonable persons" accept 19th Century science. If you doubt it, by their definition, you have lost your ability to reason. Another version is "all reputable scientists agree." By definition, any scientist who doesn't agree is not "reputable." When you have the facts, argue the facts. When you don't, attack your opponent's reasoning ability or integrity.

**Pius XII Was Only Advancing His Theory?**

> The controversy–the one over Adam and Eve–involves the competing theories of polygenism and monogenism, that is, the question of whether humans descended from many progenitors, as science argues, or from one couple, as Genesis seems to posit.

Based on the above quote it seems that professor Guinan teaches that the definitive teaching of Pius XII that polygenism is a belief that Catholics may not hold, is just a theory in competition with the opinions proposed by such as Ernst Mayr, Stephen Jay Gould, Richard Dawkins and Fr. Guinan. The article continued:

> In the past, the church's statements regarding original sin have presumed that Adam and Eve were historical people," the priest explained. "The question of monogenism and polygenism never occurred to" those writing the documents.

Assuming that by "in the past" Guinan means before the second half of the 20th Century, he is probably right. Who knew that God would reveal polygenism to atheist Ernst Mayr and his colleagues at Harvard? The CNS article continued:

> He said that the most recent statement to mention this debate is Pius XII's "Humani Generis," a 1950 encyclical. It states: "The faithful cannot embrace that opinion which maintains that Adam represents a certain number of first parents. Now it is in no way apparent how such an opinion can be reconciled with that which ... the documents of the teaching authority of the church propose with regard to original sin, which proceeds from a sin actually committed by an individual Adam."

Note to Fr. Guinan, editors of diocesan newspapers, the Catholic News Service and the Bishops who employ them: That was not merely a "statement" which mentioned "this debate." Pius XII wasn't debating. It was a definitive and binding teaching of the Ordinary Magisterium. We Catholics take the Magisterium seriously.

Without addressing *Humani Generis*, Professor Guinan moved on to his private interpretation that by an instruction from the Pontifical Biblical Commission practitioners of the historical-critical method (a.k.a. the scientific method) were authorized to reject "fundamentalist reading of Scripture" which, in what seems like Guinan's opinion, was Pius XII's sorry mistake.

> In the six decades since that document, Father Guinan continued, "the Catholic Church has accepted the use of historical-critical tools to understand the Scriptures, which are, among other things, historical documents. The 1993 instruction of the Pontifical Biblical Commission on 'The Interpretation of the Bible in the Church' calls the

historical-critical method 'essential' and rejects explicitly a fundamentalist reading of Scripture."

**Doing What Ratzinger Accused Them of Doing**

In Cardinal Ratzinger's 1988 speech at the beginning of this chapter he explained how "the model of evolution was applied to the analysis of biblical texts" beginning with the mainline Protestant Rudolf Bultman to explain "the nonhistoricity of the miracle stories." Fr. Guinan provided a sample of that in the next paragraph.

> When such an approach is applied to the Bible, he said, "Catholic scholars, along with mainstream Protestant scholars, see in the primal stories of Genesis not literal history but symbolic, metaphoric stories which express basic truths about the human condition and humans. The unity of the human race (and all of creation for that matter) derives theologically from the fact that all things and people are created in Christ and for Christ. Christology is at the center, not biology."

Probably because Fr. Guinan hasn't read anything like *Darwin's Doubt*, he has no idea that what he speculates about in the next paragraph has never been more remote and unlikely.

> He added that "the question of biological origins is a scientific one; and, if science shows that there is no evidence of monogenism and there is lots of evidence for polygenism, then a Catholic need have no problem accepting that."

In the next paragraph, the "approach" (i.e., the scientific method) in which "miracle stories" never happened but were parables for scientific method scholars to explain to us is demonstrated.

> When such an approach is followed, he said, Adam and Eve are not seen as historical people, but as important

> figures in stories that contain key lessons about the relationships of humans and their Creator.

Whatever did Christians do before Rudolf Bultman was born in 1884 to clear up the misconceptions that the Fathers, Doctors and Popes had about Adam and Eve?

> The Catechism of the Catholic Church states that "the account of the fall in Genesis ... uses figurative language, but affirms a primeval event, a deed that took place at the beginning of the history of man. Revelation gives us the certainty of faith that the whole of human history is marked by the original fault freely committed by our first parents."

Guinan, like many of those who dissent, picks a sentence from here and there to bolster his opinion. His point is that the use of "figurative language" requires scholars (like him) to interpret the meaning because it is obscure and ambiguous. Does he not know that the Catholic faith is a complete package and things have to be read in context of the entire package including Scripture and Tradition? For example, the footnote on *The Catechism* sentence that contains the phrase "figurative language" refers the reader to #13 of *Gaudium et Spes*, a document of Vatican II promulgated in 1965. *The Catechism* of 1994 says no more and no less about the use of "figurative language" than *Providentissimus Deus*, "On the Study of Holy Scripture," an encyclical issued by Pope Leo XIII in 1893 at which time Rudolf Bultman was still wearing short pants. (In the next chapter you will read Pius XII quoting Leo XIII on the proper understanding of "figurative language.) But in the next paragraph below the CNS writer gives Guinan's opinion that the Church is "straddling" the issue."

> In that language, Father Guinan detects a straddling of the issue. "It recognizes that Genesis is figurative language," he pointed out, "but it also wants to hold to historicity.

Unfortunately, you can't really have both. The catechism is clearly not the place to argue theological discussions, so whoever wrote it decided, as it were, to have it both ways."

In the next paragraph Guinan really gets weird.

> In an article about the first couple, Father Guinan wrote that Catholics who ask, "Were there an Adam and Eve?" would be better off asking another question: "Are there an Adam and Eve?" The answer, he said, "is a definite 'yes.' We find them when we look in the mirror. We are Adam, and we are Eve. … The man and woman of Genesis … are intended to represent an Everyman and Everywoman. They are paradigms, figurative equivalents, of human conduct in the face of temptation, not lessons in biology or history. The Bible is teaching religion, not science or literalistic history."

Get it, when you look in the mirror, you see the one who committed Original Sin (but he didn't actually say that, did he?)

**Another Generation**

In 2015 Stephen C. Smith had been an Associate Professor of Sacred Scripture since 2008 at Mt. St. Mary's Seminary in Maryland. Based on his biography online it seems clear that he is very intelligent and sincere. But based on his journey from Catholicism to Protestantism and back again he may have missed something regarding our Tradition. He left Catholicism as a young adult and became a leader in "one of the nation's most influential evangelical 'mega-churches' in South Barrington, Illinois." In 2000 while earning an M.A. in New Testament Theology at Protestant Wheaton College Graduate School he re-discovered his Catholic faith. He earned his Ph. D. in New Testament & Early Christianity in 2008 at Loyola of Chicago, a

Jesuit university. His doctoral dissertation was "The Determination of Criteria as Verification and Falsification Controls in the Analysis of Textual Parallels from the Jewish Wisdom Tradition and the Fourth Gospel." One can't judge a book by its cover or a dissertation by its title but that title strongly signals the historical-critical method, a.k.a, "the scientific method." "Criteria" and "Textual Analysis" are hallmarks of that method.

**It Depends on How You Look At It**

Based on a lecture Dr. Smith gave on October 6, 2015 in the Diocese of Arlington Virginia concerning *Genesis* 1-2 that was videoed and published by the Institute of Catholic Culture, it is easy to understand why some priests are not well prepared at the seminary to defend the Bible against the claims of evolution-based science that are peeling away our youth. Those leaving the Church perhaps agree with ex-Catholic celebrity Bill Maher that the Bible is a bunch of "silly stories."

Dr. Smith started his lecture with a story about his daughter being fitted for eye glasses and the way the eye doctor kept shifting the trial lenses in order to find the correct one to prescribe for her. From there he said that people get "tripped up because they don't have the right lenses on to see what God is saying." In order to understand *Genesis* 1-2 one has to see them through the correct "lenses." According to Smith there are three lenses through which people read *Genesis* 1-2: purely scientific excluding faith, literalistic which looks at Genesis with faith but not necessarily in a reasoned way and symbolically which is the correct way (according to him it seems.)

**A Visit to the Museum**

Dr. Smith spoke of some hypothetical person with whom he might visit the Smithsonian Museum of Natural History. That

Museum is totally evolutionist and has been for 100 years at least. Its evolutionist director was the one who supervised the removal of the 65,000 fossils from the Burgess Shale before 1915 (see chapter 7) and covered up their anti-Darwinian significance for decades. The Museum is also responsible for gagging Intelligent Design and shafting Richard Sternberg for daring to publish a peer-reviewed article that suggested it. (See chapter 5). Its current director is evolutionist Kirk Johnson. Dr. Smith said:

> If I went with say a Christian fundamentalist or someone who reads Genesis 1 literalistically they (sic) might have a problem if we went to the Smithsonian Museum of Natural History, right, and looked at, for example the dinosaurs, or the fossils and they might say, you know, come up with a real tension there. In fact I remember talking with a friend of mine, a very great guy, who was an Evangelical Christian a number of years ago and we were talking about paleontology and he said 'you know what happens, the reason when a paleontologist looks at a fossil that is said to be 60 million years old', he said, 'it's not actually 60 million years old.' And I said, 'It's not'? He said no its not, it's only about 6,000 years old but the Flood that God brought about in the story of Noah was so intense and so spectacular that it had this kind of, uh age, uh, age-a-fying effect on the Earth in which it makes everything look much older. And I said 'do you have a helicopter or know someone who has a helicopter because this conversation is about to get stuck in the mud.'

As he delivered that last line he paused a little but only got one titter because the people in the audience looked to me like "senior citizens" and may not have picked up on the reference to "helicopters." Black helicopters are associated with various types of conspiracy theorists such as anti-government militias. It was an insulting remark to make to a man he said was his friend and a very great guy. As I have explained elsewhere in this book, some frequently respond to creationists with *ad hominem* retorts. Possibly Dr. Smith just made up the whole story; he didn't say what that great guy replied after being equated with what many would consider "wacko extremists." Dr. Smith didn't exactly say that he believed the evolutionary dogma of the Humanists but by the way he caricatured and then derisively dismissed the opinion of one who he indicated looked at Genesis "with faith but not necessarily in a reasoned way" he gave a strong indicator that his Scripture scholarship is influenced by whatever he believes about evolution. Smith continued:

> Right, now I don't want to put him down [after he just did with the helicopter reference] but it's just to say that his particular lens allowed him to only see 6 days of creation, the world 6,000 years old because Scripture says God created the world in 6 days and on the 7th day He rested. And I would argue that while these two points of view- the purely scientific point of view and the literalistic point of view- are very, very different they have a similar problem, in fact they have the same problem. And that is that they are not reading the Scriptures, in this case, with a proper set of lenses.

I suppose that Dr. Smith thought his story was sufficient to convince his audience that anyone who believes in fiat creation and a young earth may have faith but certainly lacks reason. Dr. Smith's "lenses" don't explain his opinion that the people who look at the Scriptures from a purely scientific view and think they are just silly stories have the same problem as persons who believe that they represent historical truth. It appears to me that Dr. Smith has his own "lens" problem, namely, that he knows nothing much about evolutionary geology and evolutionary biology other than that he believes it. I wonder if the "great guy's" description of Noah's Flood as "so intense and so spectacular" is really what got the helicopter blades whirling in Smith's head because his apparent belief in an ancient Earth with 60 million year old fossils is not compatible with belief in Noah's Flood (as I explained in chapter six.)

Nothing in Professor Smith's biography indicates any work or educational expertise in natural science. As I observed in the last chapter, ever since *Humani Generis* did "not forbid" persons qualified in both science and theology to research and discuss evolution "it has been mostly theologians and philosophers telling other Catholics what they can believe about evolution based on the 'science' they learned in high school." Smith went on: "So my hope for us tonight is to make sure we have the right lenses on as we are reading Genesis 1 and 2. Sound good?" A few of the senior citizens grunted agreement.

**What Wisdom of the Church?**

He then lectured from a paper he had authored and handed out called "The Liturgy of Creation: Reading Genesis 1-2 with the Wisdom of the Church." It is not obvious how "the Wisdom of the Church" was involved because his opinions were supported by 19 footnotes, none of which were "of the Church" except for #17 which was an inconsequential reference to the CCC (Catechism of the Catholic Church): "As the Catechism reminds us, the Sabbath is the heart of Israel's law." One would think that any lecture about how to read and interpret *Genesis* 1-2 might mention *Providentissimus Deus*, the definitive encyclical "On the Study of Holy Scripture" or *Dei Verbum*, the "Dogmatic Constitution on Divine Revelation."

**A Plunge into Obscurity**

He tried to dismiss the 6- day creation narrative by noting that there are other creation texts in Scripture beside Genesis and they don't mention 6 days. (Of course, the *Psalms* are full of Creation

texts.) But to refute the understanding of *Genesis* that has come to us through Tradition (the Fathers, Doctors, Councils and Popes) Smith chose lines from the *Proverbs* (of Solomon, son of David, king of Israel) which refer to wisdom and don't mention 6 days. While *Genesis* 1 and 2 are prose narratives, *Proverbs* is not a narrative but a lot of short wise sayings. For example, here is one that certainly applies today:
*Proverbs* 29:2- When the righteous are in authority, the people rejoice; but when the wicked rule, the people groan.

The quotes from his paper that Smith read out to his audience were disjointed lines 5 chapters apart in *Proverbs* that he strung together in his paper with an ellipsis.

> 3:19-20- "The Lord by wisdom founded the earth; by understanding the heavens; by his knowledge the deeps broke forth and the clouds drop down the dew."
> 8: 22- "The Lord created me [Wisdom] at the beginning of His work, the first of His acts of old."
> 8:23 (from the King James Version) "I have been established from everlasting, from the beginning, before there ever was an earth."

"Wow" he said "It's wisdom creating and fashioning the world. Before the world was, is lady wisdom. So that leads to a question, right? Well how did it happen? Was it the 6 days of creation as in *Genesis* 1 and 2 or did lady wisdom do it all on her own somehow in one fell swoop?"

### Question Goes Unanswered

Prudently he simply left that bizarre question hanging in the air and moved on to what he said was Cardinal Ratzinger's comment on the way, in later Old Testament books,

> The Wisdom literature reworks the theme without sticking to the old images such as the 7 days. Thus we can

> see that the Bible itself constantly re-adapts its images to a continually changing way of thinking, how it changes…in order to bear witness time and again, to the one thing that has come to it, in truth, from God's word-which is the message of His creative act.

While Ratzinger didn't say what "His creative act" was, no fiat creationist has a problem with his observation about literary styles. Everyone understands the Bible is full of images and literary forms. So is the Mass. For example when Jesus advised people to cut off their hand or pluck out their eye He was using non-literal form. The "hand of God" doesn't mean that God actually has hands. Anthropomorphisms are figures of speech which represent God as having human characteristics, form or personality. They are *symbolic* descriptions, which help to make God's attributes, powers and activities real to us. (What would the ceiling of the Sistine Chapel be without them?) But how does that negate the straight-forward historical prose narrative of Genesis 1-2 especially when Jesus quoted from it when giving His definitive teaching on marriage? For an excellent, short explanation of anthropomorphisms in Genesis 1 and 2 see "Does God Have Body Parts?"
creation.com/does-god-have-body-parts

**Rug- Pulling Prof**

Then Smith explained his intellectual triumph over any of the seminarians subject to his authority who might have become 6-day fiat creationists as the result of their parents who encouraged their vocations.

> Now hear this, this really startles my seminarians when I read this next quote, 'cause for them it is pulling the rug out from under the six days of creation. Listen to this very interesting quote.

Smith then read the continuation of the Ratzinger quote that was started above but interjected his own commentary and emphasis. "In the Bible itself the images are free and they correct themselves over time. In this way they show, by means of a gradual and interactive process, that they are"-*wait for it*- "only images, which reveal something deeper and greater." Smith, gesturing like a choral director, said "Everyone say it with me, they are only images!" Some repeated "only images." He then repeated Ratzinger, "which reveal something deeper and greater." Again the choral director said, "Repeat that last part with me, which reveal something deeper and greater." When some repeated it he proclaimed that "we are already on our way to get fitted for some very nice lenses here with help from Cardinal Ratzinger."

Whatever Smith thinks the great significance of Ratzinger's observation is, the future Pope no doubt accepted the Church's teachings that all of the Bible's texts were written centuries ago under the inspiration of the Holy Spirit, are inerrant, and that nobody is producing new Bible texts to re-adapt the Bible's "images to a continually changing way of thinking" such as proposed by Hutton, Lyell, Darwin, Mayr, Gould, or Teilhard de Chardin. However, as you know from reading the first page of this chapter, Ratzinger said that in some schools "the model of evolution was applied to the analysis of biblical texts" and he considered that to be destructive to faith. In other words there are Bible scholars making up new interpretations of the ancient texts in an attempt to harmonize them with their vague notions of what they think they know about evolution.

### Genesis 1 and 2: A Mish-Mash of Symbolism

Poor Cardinal Ratzinger! After Smith established that somewhere in his voluminous writings before he became Pope he once wrote that "the Wisdom Literature" (Proverbs) contained a lot of imagery "which reveal something deeper and greater," some who attended Smith's lecture may have blamed him for encouraging the direction followed in Smith's paper. The theme of the lecture had been set in the 2nd paragraph of Smith's paper by two sentences said to be from the paragraph 327 of the Catechism of the Catholic Church (CCC):

> God Himself created the visible world in all its richness, diversity, and order. Scripture presents the work of the Creator *symbolically* as a succession of six days of divine 'work', concluded by the 'rest' of the seventh day.

The CCC text has "work" and "rest" in quotation marks to indicate those words are symbolic. Of course God doesn't literally "work" and "rest"; He is pure spirit. But when He had Moses write it, God knew how to communicate with us humans in a way that would be forever understood even by people without a Ph.D. in New Testament Theology from a Jesuit university. It seems that to Smith the whole narrative is so symbolic that he has to explain it to the uninitiated. The text of Genesis from 1:1 to 2:3 was presented in his paper as symbolism concerning Worshipping God in His Holy Temple, The Sacred Space of the Garden of Eden, The God of the Mountain, The Temple of Mt. Eden, and The Threefold Structure of the Cosmic Temple. The last 6 pages of the paper are a mish-mash of symbolism about Temples real and spiritual and the Hebrew Sabbath. I could not detect within those pages the doctrines that the Church teaches and that are derived directly from *Genesis* 1 and 2. Dr. Smith's discussion of Genesis 1 and 2, particularly with its emphasis on temple symbolism has a resemblance to the scholarship of J.H. Walton, a well-published Protestant author and Professor of Old Testament, who joined the faculty of Wheaton College around the time Smith studied there. For example, Walton's books such as *Genesis 1 As Ancient Cosmology*, *Ancient Near Eastern Thought and the Old*

*Testament*, and *The Lost World of Genesis 1: Ancient Cosmology and the Origins Debate* are heavy on creation cosmology according to ancient Egyptian and Mesopotamian thought. Walton has written that "creation texts do follow the model of temple-building texts" For example in *Genesis 1 As Ancient Cosmology* Walton uses terms such as "close association between temple and cosmos", the "seven-day temple inauguration" and the "intrinsic relation between cosmos and temple."

Whatever one may think of Dr. Smith's paper, "The Liturgy of Creation: Reading Genesis 1-2 with the Wisdom of the Church," it must be asked if his paper owes more to the wisdom of scholars like J.H. Walton than to the wisdom of our Catholic Tradition.

### Go Forth and Teach All Nations Symbolism

Although Dr. Smith did not footnote it, the quote from Cardinal Ratzinger that Smith found so significant is probably from the same book from which he had quoted Ratzinger earlier in the paper. That book is *In the Beginning...* published in 1995. It is perhaps ironic that in that book's Preface the Cardinal wrote that:

> "...the creation account is noticeably and completely absent from catechesis, preaching, and even theology. The creation narratives go unmentioned; it is asking too much to expect anyone to speak of them."

If seminary students are being taught to see Genesis 1 and 2 through the lens of symbolism rather than as the real history Church documents assert that it is, it is not surprising that some priests might have trouble standing in a pulpit and going up against the Humanist scientific consensus familiar to and believed by many, if not most, in the pews. Disarmed by symbolism, they are like the disarmed victims of mass murderers in gun-free zones; completely helpless against those whose bogus "science" is killing faith.

## A Lecture from a Seminarian

In an effort to evangelize seminarians this writer sent by email to 49 seminarians of a southern archdiocese an offer to supply to them this book free if they were interested. The offer included a link to a book review that had been published and explained what the book was about.

.

None of these future priests were interested but one was kind enough to explain to me that evolution is not a problem and that those who think it is [I plead guilty] risk giving the Church a bad name.

> Evolution in itself is not a threat. In fact, saying we cannot be open to it probably does far more damage to the faith. In the words of St. Augustine:
>
> "Usually, even a non-Christian knows something about the earth, the heavens, and the other elements of this world, about the motion and orbit of the stars and even their size and relative positions, about the predictable eclipses of the sun and moon, the cycles of the years and the seasons, about the kinds of animals, shrubs, stones, and so forth, and this knowledge he holds to as being certain from reason and experience. Now, it is a disgraceful and dangerous thing for an infidel to hear a Christian, presumably giving the meaning of Holy Scripture, talking nonsense on these topics; and we should take all means to prevent such an embarrassing situation, in which people show up vast ignorance in a Christian and laugh it to scorn. The shame is not so much that an ignorant individual is derided, but that people outside the household of the faith think our sacred writers held such opinions, and, to the great loss of those for whose salvation we toil, the writers of our Scripture are criticized and rejected as unlearned men. If they find a Christian mistaken in a field which they themselves know well and

> hear him maintaining his foolish opinions about our books, how are they going to believe those books in matters concerning the resurrection of the dead, the hope of eternal life, and the kingdom of heaven, when they think their pages are full of falsehoods on facts which they themselves have learnt from experience and the light of reason? Reckless and incompetent expounders of holy Scripture bring untold trouble and sorrow on their wiser brethren when they are caught in one of their mischievous false opinions and are taken to task by those who are not bound by the authority of our sacred books. For then, to defend their utterly foolish and obviously untrue statements, they will try to call upon Holy Scripture for proof and even recite from memory many passages which they think support their position, although "they understand neither what they say nor the things about which they make assertion.""
>
> I am not saying your intent is diabolic by any means. Indeed I think you probably have a good intent, but I believe it is in error and dangerous to the faithful. Evolution is a scientific theory that may or may not be true and you are welcome to hold it as false if prudence leads you to that conclusion, but beware of falling into the error Saint Augustine warns us of. Please humbly consider this possibility.

Observe that the seminarian stated that the theory of evolution may or may not be true and welcomed me to hold that it is false if "prudence" leads me to that conclusion. However, he cautioned me against acting on my conviction because for a Catholic to oppose evolution may subject the Church (by association) to ridicule by those who "know" it is true. Those who "know" evolution is true are informed by persons such as the late Harvard genius Stephen Jay Gould who wrote in 1982 that:

> And human beings evolved from apelike ancestors whether they did so by Darwin's proposed mechanism or by some other, yet to be discovered.

All I could reply to this young scholar was to point out that a lot of natural science research had been done since St. Augustine's day and that what he has learned as "settled science" in school is far different from the problems evolutionists admit among themselves in the peer-reviewed professional journals. I said I was sorry he was unable to accept my invitation to learn 21st Century science and what the Magisterium teaches on the subject.

He is not to blame. That's just what is taught in the seminary by the scientific-method bible scholars. He and his classmates, if they persevere to ordination, may spend the next 50 years agreeing with the Humanists that a scientific theory of human origins that may not be true can be taught to Catholics (contrary to the Magisterium) and continue to pollute understanding of the Gospel. No wonder Catholic youth are leaving and these future leaders of our church are clueless.

Many people in the church claim that consistent creation teaching "turns people away from the Gospel." However, both logic and many testimonies find exactly the opposite: *capitulation on creation* turns people off the Gospel! Conversely, consistency on creation has helped many to realize the consistency of the Gospel message.
If anyone wants to know what St. Augustine taught…
https://creation.com/images/pdfs/tj/j24_1/j24_1_5-6.pdf.
Want to know what Thomas Aquinas taught? https://creation.com/thomas-aquinas-young-earth-creationist

# Chapter 10-The Scholars' "End Run"

As noted previously, the research and discussion that Pope Pius XII did not forbid provided the guidelines were followed never happened. The priest-scholars of the scientific method, who were among those of whom he was speaking when he said they "rashly transgress this liberty of discussion" did not let up. They just kept right on lecturing and publishing books, but still seeking to eventually bring the Church around to their way of thinking. Non-Catholics don't employ them or buy their books so they have to "stay inside" the Church to make money.

Football fans know that when one's opponent has a strong defense in the center, the offense can sometimes succeed by running around it. The body of Church teaching on correct interpretation of the Bible is a solid and deep line. From the beginning of evolution-inspired attempts to remove or explain away the supernatural events on the basis of a philosophy said by its adherents to be based on "science," the Popes have responded with correction. It seems logical that if a scholar-priest spends his whole adult life in the scientific method, trying to theorize natural explanations for the supernatural events, he is going to be frustrated by the Church's insistence that the Bible is inerrant. The Second Vatican Council in the 1960s provided an opportunity for an end run. The formal documents that resulted from the Council were many. All of these resulted from drafts of proposals and working papers circulated well in advance to the world's bishops, revised and revised by comments and suggestions from bishops and their theologian consultants. Finally, when physically gathered in Rome, and supported by whatever staff they needed, bishops arrived at documents they recommended to the Pope to approve and promulgate.

## The Opportunity

One such document was *Dei Verbum*, The Dogmatic Constitution on Divine Revelation, promulgated by Pope Paul VI in 1965. Divine Revelation is the Bible and Tradition. Many of the scientific method school were advisors or consultants to bishops, and some worked on the process that resulted in *Dei Verbum*. Fr. Schillebeeckx authored papers said to have had some influence. As can be said of most documents put together by a committee, the documents of Vatican II tend to be wordy, nuanced, and, at times, sufficiently ambiguous to mean different things to different people. That works to the advantage of persons trying to bring the Church around to their way of thinking. There is a sentence in paragraph 11 of *Dei Verbum* that has been open to manipulation and bad translation from the Latin, ever since it was promulgated. The English text from the Vatican website is:

> Therefore, since everything asserted by the inspired authors or sacred writers must be held to be asserted by the Holy Spirit, it follows that the books of Scripture must be acknowledged as teaching solidly, faithfully and without error that truth which God wanted put into sacred writings for the sake of salvation. [Underline added.] Therefore "all Scripture is divinely inspired and has its use for teaching the truth and refuting error, for reformation of manners and discipline in right living, so that the man who belongs to God may be efficient and equipped for good work of every kind."

The words "for the sake of salvation" were underlined to explain the "end run" tried during a 2008 Synod of Bishops dedicated to "The Word of God in the Life and Mission of the Church." A working paper put forth and discussed during that Synod proposed that the sentence that includes the words "for the sake of salvation" should be understood and taught to mean:

> Although all parts of Sacred Scripture are divinely inspired, inerrancy applies only to "that truth which God wanted put into sacred writings for the sake of salvation."

**Discussion**

According to Brian W. Harrison, O.S., M.A., S.T.D., in an article first published in the theological journal *Divinitas* in 2009,

> This reading of *Dei Verbum*, characterized pointedly here by the restrictive words "Although" and "only," has indeed been very widespread for over four decades in Catholic faculties of theology and seminaries. Nevertheless, it was challenged by some participants in the Synod, and the Synod Fathers finally refrained from endorsing it.

According to Fr. Harrison those who proposed "Although" and "only" were arguing for "restricted biblical inerrancy—the thesis that some affirmations of the human writers of Sacred Scripture are <u>not</u> there 'for the sake of [our] salvation' and these affirmations enjoy no guarantee of inerrancy."

Fr. Harrison explained how that thesis was contrary to the constant teaching of the Church in more detail than needs to be repeated here, but to give a flavor of the arguments one example - in the 1943 encyclical *Divino Afflante Spiritu,* Pius XII quoted copiously from, and strongly confirmed, his predecessor Leo XIII. To rule out the claim that Scripture can err when it treats of certain subjects, Pius referred to what Leo said in the 1893 encyclical *Providentissimus Deus* (On the Study of Holy Scripture):

> With grave words did he proclaim that there is no error whatsoever if the sacred writer, speaking of things of the physical order, "went by what sensibly appeared," as the Angelic Doctor says, speaking either "in figurative

> language, or in terms which were commonly used at the time, and which in many instances are in daily use at this day, even among the most eminent men of science."

As noted by Fr. Harrison, Pius XII then went on to recall that his predecessor also insisted that the Bible's historical passages must likewise be defended from every charge of error. He then concluded this section of his encyclical with the following declaration, in which the thesis of restricted inerrancy is described as absolutely incompatible with "the ancient and constant faith of the Church." (The expressions in quotation marks are again citations from *Providentissimus Deus*):

> Finally, it is absolutely wrong and forbidden "either to narrow inspiration to certain passages of Holy Scripture, or to admit that the sacred writer has erred," since divine inspiration "not only is essentially incompatible with error but excludes and rejects it as absolutely and necessarily as it is impossible that God Himself, the supreme Truth, can utter that which is not true. This is the ancient and constant faith of the Church" (DS 3292-3293).

Fr. Harrison noted that although Leo XIII acknowledged the existence of *apparent* errors in Scripture, he nevertheless firmly rejected any theory of restricted inspiration or inerrancy as a supposed solution to such problems. Leo described as "intolerable"

> ... the theory of those who, in order to unburden themselves of these difficulties, have no hesitation in maintaining that divine inspiration pertains to nothing more than matters of faith and morals. This error arises from the false opinion that, when it is a question of the truth of biblical affirmations, one should not so much

> inquire into what God has said, but rather, into why He has said it.

Fr. Harrison explained why this is important to our faith and why those scholar-priests, who specialize in interpreting the Bible according to naturalism, prefer, contrary to Leo, to reinterpret "what" God said so they can tell us why He has said it:

> This illicit question as to "why" rather than "what" would in practice be very frequently invited by the proposition that we are criticizing. For when faced with any seemingly erroneous statement of a biblical author, the apologist or Scripture scholar who follows the [proposed interpretation] teaching will inevitably be led to ask the obvious "why" question: "Is this statement here *for the sake of our salvation*, or not?" And if he can persuade himself that the problematic biblical affirmation is *not* salvific in purpose (as he almost certainly will when it is about history or the physical cosmos), then he will complacently dispense himself from the task of having to defend its truth. For the [scientific method] school of thought reassures him that biblical authors can in fact perform the remarkable feat of penning statements that are erroneous and yet divinely inspired.

Fr. Harrison, in mentioning "history or the physical cosmos," is naming the one area of Scripture, *Genesis* 1-11, that some scholar-priests have made a particular target. In this and previous chapters, the activity of some scholar-priests, their methods, and their inclinations to harmonize Catholic belief with naturalism and evolutionary philosophy have been considered. Who can judge their hearts, but one can speculate on possible reasons why they acted thus. Perhaps the scholar-priests believe that proved scientific facts demand it or they lack the faith in the Church's teaching, as so many did and still do with regard to *Humanae*

*Vitae*, to go against the opinion of the secular consensus. Then there is the praise and fame that the world gives to all Catholics who dispute Church teachings.

### A Current and Widespread Error

A modern American testimony that was published on the blog UnamSanctamCatholicam.com confirms that teachers, not necessarily through malice, but through ignorance, spread false interpretations of *Dei Verbum* that were passed to them by their teachers. The below testimony is extremely relevant because it concerns a relatively new Catholic college that enjoys a fine reputation for orthodoxy. The title of the article is "Inspiration 'for the sake of our salvation.'"

> Let's talk about Dei Verbum 11. Few Conciliar documents give me more headaches than this one passage out of the Constitution on Divine Revelation. The passage states that the Bible "teaches, without error that truth which God wanted put into the sacred writings for the sake of our salvation." As we know, this passage is universally misapplied by modern Scripture scholars to mean that only those things pertaining to salvation can be considered to be truly inspired. Nor is this interpretation made by liberal or modernist scholars either; otherwise orthodox Catholic Scripture scholars read the document the same way. Back when I was at [name withheld], our professor of Sacred Scripture ...had us read Dei Verbum and told us that only those parts of the Scriptures that pertained to faith and morals could be considered inspired, and therefore infallible. When I objected and stated that he was misinterpreting Dei Verbum 11, he looked at me blankly and said that he was "not aware of any other interpretation."

The former student attended before 2007, but confirmed to this writer in February 2015 that the teacher was still employed there.

# Chapter 11-Does Truth Matter?

"Post-truth" was declared by the Oxford English Dictionary to be their 2016 international word of the year. It is defined as, "relating to or denoting circumstances in which objective facts are less influential in shaping public opinion than appeals to emotion and personal belief." This, of course, is really a continuation of post-modernism which denies the existence of truth as an objective, absolute reality. Who can say if accepting the scientific consensus and arbitrarily inserting God somewhere into the process is detrimental to any particular student's Faith? Although relativism is the hallmark of this era, some Catholic students are taught that "truth matters," and that "there can be no conflict between science and religion because God is the Author of both." By overlaying God's supernatural role on the naturalist consensus of evolution they can arrive at what is, for them, both scientific and religious truth. But is it actually true? I explain in this chapter a hypothesis regarding the effect that holding a "scientific truth" that is integral to Humanist philosophy and its worldview, while also holding a "religious truth" that is rejected by the scientific consensus may have on such individuals.

The first point that cries out to be made is that if theistic evolution "works" for many Catholics, it doesn't "work" for everyone. Earlier I mentioned a friend who believed that life from non-life had been demonstrated. That friend left a private Catholic high school as a theistic evolutionist. While earning undergraduate and graduate degrees in science, doubt entered. After reading Richard Dawkins's *The God Delusion* (8.5 million copies sold), all belief in God vanished. Anecdotes are proof of nothing but the accelerating loss to the Church of young adult members, proved by Pew research and the CARA studies, requires at least a working hypothesis to explain the growing exodus of younger Catholics.

### How Some Teens Resolve the Conflict

In the first place, to be taught each part of the theistic evolution combination separately (naturalistic science in school, supernatural religion at church and home) can lead to "cognitive dissonance." Cognitive dissonance is defined as a situation involving *conflicting* attitudes, beliefs, or behaviors. This produces a *feeling of discomfort,* leading to an alteration in one of the attitudes, beliefs, or behaviors to reduce the discomfort and restore balance. At some point, teenagers recognize that the naturalistic evolutionary model of origins and the supernatural, fiat creation model described in the Bible, which they have heard read at Mass even if they never opened a Bible, can't both be true. They experience cognitive dissonance and to relieve the conflict they must alter their beliefs in one direction or the other.

The evolutionary scientific consensus has been taught to them five days a week, year in and year out, as proven science. They incorrectly associate that evolutionary, pre-historic "science" with the real science (engineering really) that delivers the goods including all of their favorite electronic gadgets. The science they have been taught provides a coherent explanation of the origin of all reality. The alternative to the scientific consensus that they have been taught to believe is that a loving, personal God, created the universe and all that is in it out of nothing by an act of His Will. Further, His Body, Blood, Soul and Divinity are physically present in that locked, gilded box in the church. Some might find it easier to believe in science. A teenager may have been taught what the Church teaches, namely that the entire Bible is the Word of God, written by human authors under the inspiration of the Holy Spirit. Doesn't the belief that the supernatural transubstantiation of bread into the Body/Person of Jesus when the priest says "This is My Body" depend on the Bible's account that a supernatural event took place at the Last Supper and that the power to repeat that supernatural event was delegated to men through the Church is literally true? A teenager may ask: "But if the beginning of the Bible, that is, the account of our supernatural origins, should be interpreted allegorically to accommodate the claims of naturalism, why should other parts of it, such as supernatural transubstantiation, be interpreted literally?" The biblical scholars Cardinal Ratzinger criticized in the speech quoted in Chapter Nine have answered that question: "It shouldn't! We have to explain how those miracle stories came to be made up."

## Numbing Effect on the Faith of Youth

In his 2007 book, *The Doctrines of Genesis 1-11: A Compendium and Defense of Traditional Catholic Theology on Origins*, Fr. Warkulwiz made an enormous contribution to the Church. One of his observations seems appropriate to repeat here while considering the effect of conflicting information on the youth.

> The theory of evolution has caused confusion in the minds of the young because it differs so much from what is in the Bible. They recognize the contradictions and are not sophisticated enough to rationalize them away. Pope Leo XIII said in *Providentissimus Deus*: "[F]or the young, if they lose their reverence for the Holy Scripture on one or more points, are easily led to giving up believing in it altogether. It need not be pointed out how the nature of science, just as it is so admirably adapted to show forth the glory of the Great Creator, provided it be taught as it should be, so, if it be perversely imparted to the youthful intelligence, it may prove most fatal in destroying the principles of true philosophy and in the corruption of morality."

> The notion of an earth billions of years old, which is espoused by Catholic evolutionists, has had a numbing effect on the faith of youth. It pushes God so far into the background of time that He's barely visible and hardly seems relevant today. But the God of *Genesis* is up front. He created the world only a few thousand years ago and has lovingly and providentially followed, and intervened in, the history of mankind.

"What," students may wonder, "was God doing during those billions of years?" Especially in his public school education, the student will be exposed to the concept that religion is just a cultural development of the pre-scientific age. To resolve that cognitive dissonance caused by those questions some might decide to reject the Bible outright (and belief in God as we know Him). And many have. In Chapter One this writer outlined the Pew Research showing how many have. John West of the Discovery Institute published an article titled "Are Young People Losing Their Faith Because of Science?" in which he observed that his research

> suggests that if churches want to be effective in answering student questions about science and faith, they cannot wait until adulthood or rely on college ministries to do the job. They need to be engaging young people on these issues when they are in middle school and high school, if not earlier.

Along that line of thinking is an article called "Creation—the 'dealbreaker' for today's generation: Survey results surprise youth worker" which was written about Australia but applies equally to America. Mentioned is a survey of religious beliefs of Generation Y (born after 1980), which showed that less than 50% of that group even believed that there was *any* sort of God, and that the *single biggest* reason that this generation gave for loss of faith was "doing further study, **especially science.**" (Mason, M, Singleton, A., and Webber, R., *The Spirit of Generation Y)*. The article is online at www.creation.com/creation-dealbreaker

What those who have studied the loss of our youth have determined is that the "origins" questions raised by Humanist "science" must be countered very early. This means it is the duty of priests and parents. Priests have the control of parish facilities and can use them to promote and facilitate truth in natural science education. If priests lead, parents will follow.

### The Specific Educational Need

Indoctrination in evolution is a contributing factor to the alienation of Catholic youth that can, and should be, countered at the parish level by the mutual cooperation of parents and clergy to first educate themselves on critical matters which are now "below their radar," so to speak. A string of Popes have been addressing the problem of naturalism/evolutionism in formal documents of the Ordinary Magisterium since at least 1893. But to the detriment of the Catholic people, their theology and philosophy elites turned a deaf ear, and as a result, many of our clergy and people have lost confidence in some basic truths of the Faith. In doing that, they also lost confidence in the Catholic worldview and surrendered public policy to the Humanists. Stemming the tide of Catholic loss depends on action at the parish level, which is where the children are at the age when most Catholics who do lose their faith, lose it.

What if during that struggle the teen had someone to explain with confidence that evolution is primarily speculation held together with circular reasoning that interprets all observable data based on the premise that evolution is unquestionably true, does not need to be proved, and merely needs to have theoretical guesswork to explain it? Remember that any evolution-supporting theory at all, no matter how utterly implausible, is considered "science." If the reader thinks this writer has mischaracterized the premise of the evolutionists, consider the aforementioned 318-page book *What Evolution Is* by the "Darwin of the 20th Century." atheist Ernst Mayr. In its Preface he wrote:

> Also there is no longer any need to present an exhaustive list of proofs. That evolution has taken place is so well established that such a detailed presentation of the evidence is no longer needed. In any case, it would not convince those who do not want to be persuaded.

That sounds so superior and confident, but the reality is that it is the overt policy of the scientific consensus to not engage in two-sided debate with scientists who support either the creation model or intelligent design. University icons like Mayr write books and pontificate in their own closed circle. They are never told "they are naked" because there are no little boys such as the one in the tale, "The Emperor's New Clothes," to tell them.

### Evolutionists Avoid Live Debates

The scientific consensus was furious when in 2014, Bill Nye "the Science Guy" who helped propagate evolution to generations of school children via his PBS TV programs, accepted a live debate with Evangelical Ken Ham. Ham is founder of the creationist organization, Answers in Genesis. The live debate was to be streamed on the internet. A few days before the debate there was an article in the *Washington Post* by a writer for the Religion News Service, "Ham-on-Nye debate pits atheists, creationists," that captured the evolutionists' angst:

> "Scientists should not debate creationists. Period," wrote Dan Arel on the Richard Dawkins Foundation's website. "There is nothing to debate." Arel, a secular advocate, is echoing the position of Dawkins, an evolutionary biologist and outspoken atheist who has long refused to

> debate creationists. "Winning is not what the creationists realistically aspire to," Dawkins said in 2006. "For them, it is sufficient that the debate happens at all. They need the publicity. We don't. To the gullible public which is their natural constituency, it is enough that their man is seen sharing a platform with a real scientist."

### Humanist Attack on Nye

Ken Ham is an intelligent man with an undergraduate degree in science who once taught science in high school, but he is not a top-tier scientist or a seasoned live debater like the Ph.D.s from the Institute for Creation Research that the Humanist consensus fears most. With his years of experience telling evolutionary stories as a professional performer on TV, it should have been a "cakewalk" for Bill Nye. It wasn't, and the Humanists were on him like wolves. A post-debate article in the online publication, *The Daily Beast*, popular with those of Humanist persuasion and the left wing of the American political divide was headlined, "The Bill Nye-Ken Ham Debate Was a Nightmare for Science." The sub-headline was, "In a much-hyped showdown, 'the Science Guy' tried to defend evolution against creationist Ken Ham, and proved how slick science deniers can be. How did the guy who's right go so wrong?" The writer was so angry with "the guy who is right" that he launched a rather vicious attack on Nye's integrity. After suggesting that Nye might have been bribed to go to Ham's Kentucky headquarters for the debate he added:

> Actually, there are two other reasons that Nye might have done so, and I've given both possibilities a great deal of thought in the past few days. The first is that Nye, for all his bow-tied charm, is at heart a publicity-hungry cynic, eager to reestablish the national reputation he once had as the host of a PBS show. When his stint on Dancing With the Stars ended quickly, Nye turned to the only other

> channel that could launch him back to national attention: a sensationalized debate, replete with the media buzz that he craves.

The author of that article, Michael Schulson, holds a B.A. in Religious Studies from Yale. One could easily infer that with that background he is trained in Humanist philosophy, the "scientific method" of interpreting Scripture, and that the only thing he really knows about evolution is the same thing that atheist Dr. Ernst Mayr wrote in the Preface to his book quoted earlier to the effect that evolution is so well established in the minds of the Humanists that there is no need to present evidence. For that reason it is unlikely that Mr. Schulson has any idea how hard it is to debate with someone like Mr. Ham who knows his subject because it is even more unlikely that Mr. Schulson ever heard anything but evolutionism during his time at Yale. He could think of no reason why his "Science Guy" fared so badly so he attacked his character and motivation. Nevertheless, Mr. Schulson was able to assure *The Daily Beast's* readers that "what Ham was saying made absolutely no scientific sense." With his Ivy League education Schulson could only bluster that Ham is a "science denier" and that his argument is "bulls _ _ t."

### Humanists Prefer the Opinion of Humanist Judges

Earlier, in discussing Fr. Vawter, it was noted that the information about him came from a deposition in connection with a court case. The case was to overturn the Arkansas *Balanced Treatment for Creation-Science and Evolution Science Act.* That Act, passed by an elected legislature, stated that "Public schools within this State shall give balanced treatment to creation-science and to evolution science." If evolutionists are so confident in their science, would they not be able to quickly prove to the students that creation science was, to use a good Ivy League expression, bulls_ _t? They should welcome the opportunity to put that "religion passing itself off as science" in its place pretty quickly in head-to-head comparison one would think. However, they took the ACLU-New York law firm approach. Evolutionists seek contests in the courtroom and fear contests in the classroom because when the creation/intelligent design model of interpretation and the evolution model of interpretation are applied to the same data, the creation model or the ID model is more plausible to anyone who will admit that God might exist.

## Liar, Liar Pants on Fire

Arkansas, and later Louisiana, lost in Court as judges cited the testimony of witnesses that evolutionism is science and creation-supporting science is just religion. In the Arkansas case, the judge said evolution was science because it was open to revising its theories but creation-supporting science was religion because it had already reached its conclusion. One of the key witnesses produced by the ACLU lawyers in the 1982 Arkansas case was Michael Ruse who was a professor of philosophy and zoology at the University of Guelph, Canada. His testimony loftily dismissed the claim that evolution was an anti-god religion. However, based on an article he wrote for the *National Post*, May 13, 2000, called "How evolution became a religion: creationists correct?", Professor Michael Ruse knew that in fact evolution is an anti-God religion even as he denied it under oath in the Arkansas court case. He wrote:

> Evolution is promoted by its practitioners as more than mere science. Evolution is promulgated as an ideology, a secular religion—a full-fledged alternative to Christianity, with meaning and morality. I am an ardent evolutionist and an ex-Christian, but I must admit that in this one complaint—and Mr. Gish [the late Dr. Duane Gish of the Institute for Creation Research] is but one of many to make it—the literalists are absolutely right. Evolution is a religion. This was true of evolution in the beginning, and

it is true of evolution still today. … Evolution therefore came into being as a kind of secular ideology, an explicit substitute for Christianity.

### Evolutionists Know Evolution is a Religion

Richard Lewontin (b. 1929), is another of the New York city-born atheist sons of Eastern European Jews who became famous evolutionists. Carl Sagan and Stephen Jay Gould were more famous because they were media personalities but Lewontin was a more "heavyweight" academic. Dr. Lewontin's field is evolutionary biology and evolutionary population genetics which in 1966 he combined into evolutionary molecular genetics. Lewontin held an endowed chair in zoology and biology at Harvard for 25 years. He collaborated with atheists Gould and Ernst Mayr who were also at Harvard during most of those years. On January 9, 1997, *The New York Review* [of Books] published a review of Carl Sagan's *The Demon-Haunted World: Science as a Candle in the Dark.* The review article was "Billions and billions of demons" and it was written by Professor Lewontin. The words in italics were in italics in Lewontin's original:

> Our willingness to accept scientific claims that are against common sense is the key to an understanding of the real struggle between science and the supernatural. We take the side of science in spite of the patent absurdity of some of its constructs, *in spite* of its failure to fulfill many of its extravagant promises of health and life, *in spite* of the tolerance of the scientific community for unsubstantiated just-so stories, because we have a prior commitment, a commitment to materialism.
>
> It is not that the methods and institutions of science somehow compel us to accept a material explanation of the phenomenal world, but, on the contrary, that we are forced by our *a priori* adherence to material causes to create an apparatus of investigation and a set of concepts

> that produce material explanations, no matter how counter-intuitive, no matter how mystifying to the uninitiated. Moreover, that materialism is absolute, for we cannot allow a Divine Foot in the door. The eminent Kant scholar Lewis Beck used to say that anyone who could believe in God could believe in anything. To appeal to an omnipotent deity is to allow that at any moment the regularities of nature may be ruptured, that Miracles may happen.

### Your Children Taught Their Religion

In 1884 Pope Leo XIII wrote the encyclical *Humanum Genus* (On Freemasonry and Naturalism). This will be quoted from extensively in Chapter 14. In it he explained how Naturalists and Freemasons would use the schools to integrate evolution into philosophy to the detriment of morality and culture. That the schools have been and continue to be used for that purpose and with that result is a key thesis of this book. Many church-going Catholics sense the decline of morality and culture but don't see the connection. Some clergy, in their homilies, attribute the moral decline to "the work of the devil." Well even the devil needs organized help. As just one example of organized anti-theism and doctrinaire naturalism consider the National Center for Science Education (NCSE). From its name the NCSE sounds as if it is a public-interest "think tank" interested in promoting science education. In reality it is a Humanist front group formed to contend with antievolutionists and ensure that public school children never learn of evolutionary fallacies. The NCSE gets plenty of cooperation (wittingly or unwittingly) in support of its mission from Catholic schools and other Catholic educators.

### Humanist Superstar

From 1987 through 2013 NCSE's Executive Director was Eugenie C. Scott, Ph. D., a highly-decorated Humanist "superstar." Scott traveled the typical road from one religion (Christian Scientist) to another religion (Congregationalist) to another religion (Humanism) aided by her academic field of study which was the "biological and behavioral aspects of human beings, their related non-human primates and their extinct hominid ancestors." In 1980, Scott was at the forefront of an attempt to prevent creationism from being taught in the public schools of Lexington, Ky. From this grassroots effort in Kentucky and other states, the NSCE was formed in 1981. Readers are urged to read Scott's biography online at Wikipedia to see how doctrinaire evolutionists' organizations are interwoven and mutually supportive by seeing the organizations to which she belongs or which have honored her. For her life-long work of promoting evolution as the only acceptable explanation of everything she has been honored by: National Advisory Council of Americans United for Separation of Church and State; the Skeptics Society; California Academy of Sciences; American Association for the Advancement of Science; American Society for Cell Biology; American Institute of Biological Sciences; National Association of Biology Teachers; Hugh M. Hefner First Amendment Awards; Society for the Study of Evolution; National Academy of Sciences; National Science Teachers Association; American Humanist Association.

### A Gospel Openly Preached

The monthly magazine *Scientific American* is one of the pulpits from which Eugenie Scott has preached her religion. In an October 2013 article titled "Climate in the Classroom" Scott explained the successful effort of organized Humanists to censor any alternative to evolution, increase the teaching of evolution and brainwash kids on global warming hysteria.

> For decades objections to the theory of evolution have bedeviled individual teachers, school boards of education and state legislatures. Educators fought to keep evolution

in science classes and creationism out. We resisted intelligent design, the notion that natural selection alone cannot explain the complexity of life-forms, which served as a way of getting creationism through the back door. We are now fighting legislation that encourages teachers to teach 'evidence against evolution'—facts found only in creationist literature.

> The consequences of antievolutionism are felt in many American schools: evolution is not taught or is taught poorly. Yet evolution is one of the most important ideas in human intellectual history, and students have a right to learn it. The common ancestry of living things and the mechanisms of inheritance explain why things are the way they are. Students and adults deprived of this knowledge are scientifically illiterate and ill-prepared for life in a global, competitive world. Students given merely once-over light instruction in evolution are woefully undereducated...The newly-released Next Generation Science Standards, developed by a consortium that includes the National Academy of Science, 26 states and the non-profit organization Achieve, will require teachers in states adopting them to teach both evolution and climate change... Beginning learners have a right to know what scientists have concluded. It is not right to allow religious, political or economic ideologies to trump instruction in science.

Evolutionists are also climate change alarmists because they believe the geological history of the world as dreamed up by Lyell in the 19th Century. They reject Noah's Flood which caused the most dramatic climate change event in history including one ice age. Long-age interpretations of earth history have led uniformitarian climate scientists to conclude that at least 5 dramatic climate fluctuations (ice ages) occurred in the past and could occur in the future, with possibly disastrous consequences. The belief in multiple ice age is based on speculative astronomy (the Milankovitch theory), chemical wiggles called oxygen isotope ratios within two deep sea sediment cores from the southern Indian Ocean and an out-of-date 1976 paper titled "Variation in the Earth's Orbit: Pacemaker of the Ice Ages."

Climate change and evolution are now "joined at the hip" in the Humanist ideology that they require be taught. Children in schools are being indoctrinated. Humanists called for "climate-change deniers" to be prosecuted under the Racketeer Influenced and Corrupt Organizations Act.
http://dailycaller.com/2015/09/17/scientists-ask-obama-to-prosecute-global-warming-skeptics/
Democrat Attorneys General from 17 states joined forces to coordinate investigations of individuals who express skepticism toward the idea of man-made global warming, or climate change. With his "climate change" opinions in the 2015 encyclical, *Laudato Si*, it appears that Pope Francis has aligned himself with coercive ideology. EWTN broadcaster Raymond Arroyo, speaking on the Laura Ingram radio show on May 24, 2017, said that EPA officials of the Obama Administration had a hand in that which Francis published for our moral guidance. The U. S. Bishops employ a large staff in Washington called the Catholic Climate Covenant (CCC) to promote the Humanist-backed "global agreement" in the U.S., which the Bishops' executive director calls a "polluting country." The CCC has been rocked by scandal including a prominent staffer who was "outed" by the Lepanto Institute for her pro-homosexual and pro-abortion advocacy. To get the truth about climate change issues see https://realclimatescience.com/ .

Such is the state of education today, Humanists do not have to have children; they reproduce themselves by taking yours. The

*NY Times* of September 2, 2013, in a special issue devoted to science and math education published an adoring article complete with a photograph of Eugenie Scott and three skulls. According to the article Scott realized in 1974 that creationism is "a movement that could have really serious consequences for science and science education." The article explained that the NCSE "despite a relatively skimpy budget has had an outsize impact on the battles in courtrooms and classrooms over whether creationism — the idea that the universe was devised as it is by a supernatural agent — or its ideological cousin, 'intelligent design,' should be taught in public schools."

As Pew Research has demonstrated, 65% of Americans already accept evolution as a fact but Humanists aren't satisfied with that. Contrast the organized tactical and strategic evangelization by Humanists with the laissez faire approach devised by Church officials for "The New Evangelization" based on layman to layman personal communication and good example. Meanwhile the organized resources of the Bishops are devoted to "bigger issues" like climate change, illegal immigrants, and settling Muslims refugees among us. What if, instead of pouring pew-sitters' money into lobbying for a bigger coercive welfare state and more funding to import people who call us infidels and have been at war with us since the 8th century, the Bishops supported creation-supporting advocacy groups just as Humanists support creation-censoring groups? Wouldn't that help the children of the pew-sitters keep their parents' Faith? What if.

## Science Guy's Friends Rally Around

Even though Bill Nye was thrashed in the February 2014 debate with Ken Ham and had his integrity attacked by an evolutionist in *The Daily Beast* who was stunned at his ineptitude, the "superstar" evolutionists rallied around him and sought to rehabilitate him. Responding to his February disgrace, Nye had a collaborator/ editor help him to rush into print by November 2014 a book called *Undeniable: Evolution and the Science of Creation* whose front and back covers are dominated by the bow-tied science guy's close up picture. It's a silly compilation of evolution stories told in the first person. It contributed nothing to science except more propaganda for the kids who recognize his picture from watching him on PBS. However, the back cover of the dust jacket included glowing endorsements of the book bound to have it leaping off the shelves at $25.99 and into one's local public library (where I found it in hardback and paperback.) The endorsers were 5 prominent evolutionists including Eugenie Scott, Neil Degrasse Tyson (host-narrator of the 13-part 2014 Cosmos TV program of evolutionary fiction discussed in chapter 14) and Michael Shermer. Shermer is founder of the Skeptic Society and editor of its magazine *Skeptic*. He produced and co-hosted the 13-hour Fox TV series *Exploring the Unknown* (1999). He describes himself as an advocate for Humanist philosophy. His endorsement was so over-the-top it should have embarrassed him. According to Shermer:

> Bill Nye has penned one of the clearest and most moving explanations of evolution since Darwin's *On the Origin of Species*. With clarity and passion, Nye brings evolutionary theory to life.

Oh really? Nye is a TV performer. In his debate with Ham he showed he actually knows nothing. Yet Shermer credits him with writing one of the clearest explanations of evolution since 1859. Evolutionists TV stars know how to stick together. And they never shy away from preaching their religion with conviction. Yet Shermer, like Nye, knows nothing. Watch
https://www.youtube.com/watch?v=Y5tEAINU3wc

*Smithsonian*, an evolutionist magazine, in its January 2015 edition, noted that *Undeniable* was the previous autumn's surprise best seller in which "TV's 'science guy' explains why evolution is true." The article consists of "heavy scientific" questions and "intelligent" scientific answers that one should expect from the one who wrote the clearest explanation of evolution since 1859. For example:

> Smithsonian: What do you think was the most transformational moment in human evolution?
> Nye: No one is sure what happened when we got speech, when we were able to communicate with language. And that sure made a heck of a difference.
> Smithsonian: With such remarkable evidence, why do you think people have so much trouble accepting evolution?
> Nye: Most people can't imagine how much time has passed. The concept of deep time, it's just amazing.
> Smithsonian: It has been 90 years since the Scopes Trial, but many Americans don't believe in evolution. Will we still be debating it in a century?
> Nye: I think there will always be religious fundamentalists who have trouble accepting evolution. That said, I'm confident there will be a lot fewer of them. In 100 years we won't have much of it. In the next 50, there will be plenty.
> Smithsonian: Has your taste in neckties evolved?
> Nye: Yes, I like them narrower now. I prefer ties of finer fabric. In other words, I can afford nicer ties.

So there you have it. If you were not edified and educated by the "science guy's" eruditeness you must be one of those religious fundamentalists. There may be plenty of you over the next 50 years because it has taken from 1859 until now to convert 68% of the white, non-Hispanic Catholics but Humanists have the schools and they have plans for your children and your grandchildren. As Nye said, the concept of deep time is amazing. If time can turn goo into you, what's another 100 years or so to turn a religious fundamentalist into a religious Humanist? Nye mentions "when we got speech" as if it just "evolved." Tom Wolfe, the maestro storyteller and reporter provocatively argues in his August 2016 book, *The Kingdom of Speech* that what evolutionists think they know about speech evolution is wrong.

**Ordinary Catholic Adults Can Help Younger Catholics**

Any Catholic can inform himself on the Church's teaching and the scientific case against evolution to help the young accept the Church's teaching. It is not one of the objectives of this book to present the scientific case in other than layman's terms. Resources to enable one to delve deeply into the scientific case against evolution are very plentiful and many of those are free online. Some recommended sources are scattered throughout the book and summarized in Appendix III. This writer's spouse provides an excellent example of what an ordinary layman or priest can do to inform himself. In the Preface to the 2009 book she published, *A Bird in the Hand...Some Thoughts Concerning Evolution, Creation and the Teaching of the Catholic Church,* she described our journey to understanding this way:

> One day many years ago my husband was taking a seat on a bench in a park in Washington, D.C., when he noticed a small book someone had left there. He read it and it awakened his interest in the topic of the theory of evolution. He began to send away for materials from the Institute for Creation Research and to read whatever he could get his hands on. He began to realize that the theory of evolution was just that—a theory and that it had not been proven. He also learned that there was a tremendous amount of evidence available to support a young earth (10,000 or so—not billions of years) and the global flood described in the Book of Genesis. As a scientist (an engineer) himself, he became impressed with these facts and moved from being a more or less passive theistic

evolutionist to being a committed, active creationist. I went through this process somewhat second hand, since during this time I was busy having six babies and trying to keep up with the active daily life which that entailed. Then the time came when I had to make a decision. I went to a parent- teacher meeting for my oldest child, by now in seventh grade. My husband had asked me to talk to the headmaster of the school about the evolutionary material that was so evident in the science books the school used. I said to Brother Kennedy, 'My husband asked me to speak to you about the textbooks you use and to point out to you that they teach evolutionary theory as fact.' Brother Kennedy replied, 'What do you think about this topic, Mrs. McFadden?' I had no real reply, except to say I wasn't sure. That caused me to begin my in-depth education into this whole debate. I began by reading Henry J. Morris' book, *The Genesis Flood,* which convinced me that there was, truly, a worldwide flood at the time of Noah and that science has evidence to support the Bible on this issue. I began to read more and more on the topic and became, like my husband, a committed defender of what I came to recognize as the traditional Catholic doctrine of creation. And so I have remained. Recently my husband picked up a book written by a Catholic theologian who teaches at a Catholic seminary and was annoyed to find that this theologian believed that evolutionary theory is 'probable' and he used the ... historical critical method in discussing the first 11 chapters of the book of Genesis in the Bible. We stayed up late that night while he vented his frustration at this book and I suggested he write a book in response. He declined my suggestion, so I said that I would attempt to write such a book. He gave me his encouragement and what you see here is the result.

In the testimony just related, it was noted that *The Genesis Flood* contained observed data that is compatible with the world-wide flood described in *Genesis* 6 and sometimes called "Noah's Flood." The publication of that book in 1961 started the creation-supporting science revival in America and led to the founding of the Institute for Creation Research in 1970.

### Evangelicals Did the Research, Doing the Teaching

Evangelicals did the scientific research that Pius XII authorized Catholics to do but the controlling Catholic scholar-priests and lay intellectuals either weren't interested or lacked the education to do so. On the other hand, the Institute for Creation Research (ICR) has assembled a powerful group with science Ph.D.s from secular universities and other scholars and researchers who are expert in all realms of natural science, historical biblical languages and archeology. An answer to almost any scientific question related to evolution or creation can be found online at ICR.org by using the search tool on the left of the homepage. ICR's excellent monthly *Acts & Facts* is sent free.

### Evangelicals Work in Background

Earlier in this book readers were told about the Discovery Institute and urged to explore its website and get its free publications. Although the intelligent design work of that Institute remains silent on the ultimate source of the intelligent design it appears that Evangelicals are prominent in that organization. For example, Dr. Stephen Meyer who heads the Center for Science and Culture confesses his Christian belief. It appears also that it is Evangelical pastors who are inviting the Center to provide educational seminars and promoting the Center's scientific literature for the benefit of their members and members' children. For example there was a conference in November 2014 held by Evangelicals on the subject of apologetics that included Dr. Meyer teaching the evidence for intelligent design. The Evangelical's National Religious Broadcasters Association showed video of his presentations from that Conference on their NRB Network. To get just some idea about the effort Evangelical pastors are making to counter bogus evolution in their churches, look at the calendar of the Creation Ministry International's speakers crisscrossing the U.S. http://creation.com/calendar?utm_media=email&utm_content=us Another really interesting website that is sponsored by the Discovery Institute is evolutionnews.org. It has volumes of up-to-date news and analysis under the topics of evolution, intelligent design, science, academic freedom, culture & ethics, education, and faith & science. It is free.

**Theistic Evolution Powerhouses**

As I use the term "Evangelicals" in this book, I mean Protestant Christians who are fiat creationists who believe the bible is inerrant and historical as the Fathers, Doctors, Councils and Popes taught authoritatively. The majority of Protestant Christians publishing in this field are not Evangelicals but are theistic evolutionists of various stripes whose views of God, Christianity, Christ, and the bible are as varied as it gets. For example, the Templeton Foundation with a $1.5 billion endowment gives away about $70 million a year to promote theistic evolution. In a 2006 book, *The Language of God*, Francis Collins proposed "Biologos" as the new term for theistic evolution and in 2007 founded an organization of that name. Among the principal beneficiaries of the Templeton Foundation are theistic evolutionist organizations such as Biologos which promotes bizarre science and theology but with a budget estimated at $9 million it is very influential. Like most issues in life, a lot can be learned by following the money. If Foundations or Government agencies are funding a particular view there will be more scientists becoming true believers in that view as they scramble for grants and contracts.

It is beyond the scope of this book to critique in any detail the prominent authors who promote theistic evolution but an excellent critique can be found in *Aquinas and Evolution.* Fr. Chaberek did to the philosophic and theological arguments of "Thomistic" evolutionists what Dr. Stephen Meyer did to the scientific arguments of scientific evolutionists in his *Darwin's Doubt.* Both authors quote the best arguments of the evolutionists from their published work and then systematically point out contradictions and other errors. Also, *The Shadow of Oz: Theistic Evolution and the Absent God* by Wayne D. Rossiter is a critique of theistic evolution by a former atheist with a Ph. D. in biology from Rutgers who through some miracle of grace had a private late-night epiphany regarding the implications of his atheism. He shows that theistic evolution is bad philosophy and outdated science.

The Templeton Foundation and its web of theistic evolution grantees is a major opponent of the creationists and the Intelligent Design Movement and it promotes evolution among Catholics. For example, according to Businessweek.com, in 2005 the Foundation said it was one of the "principal critics" of the intelligent design movement and funded projects that challenged it. According to a report by the Associated Press, in March 2009 the Discovery Institute accused the Templeton Foundation of blocking its involvement in Biological Evolution: Facts and Theories, a Pontifical Council for Culture-backed, Templeton-funded conference in Rome. What Catholics hosted that event? The conference was held at the Pontifical Gregorian University, a Jesuit institution. One can only speculate but it is reasonable to assume that many professors and seminarians from the other colleges in Rome attended as observers. Those American seminarians that went on to ordination and persevered in their vocation are among today's clergy serving in America.

## Jesuit Cult Theology

On the lack of involvement of any speakers supporting intelligent design, the conference director Jesuit Rev. Marc Leclerc said, "We think that it's not a scientific perspective, nor a theological or philosophical one...This makes a dialogue difficult, maybe impossible."

The exclusion of Intelligence from the biological evolution of humans (even if you believe in biological evolution) doesn't leave room for God as we ordinary Catholics know Him. But Jesuits are no ordinary Catholics. Many subscribe to the theology fiction of Jesuit Teilhard de Chardin whose guide to God and the world was not Scripture or Tradition or the Magisterium or even genuine science. In his *Phenomenon of Man* de Chardin wrote:

> Is evolution a theory, a system or a hypothesis? It is much more; it is a general condition to which all theories, all hypotheses, all systems must bow and which they must satisfy henceforward if they are to be thinkable and true. Evolution is a light illuminating all facts, a curve that all lines must follow.

Teilhard is an exemplar of many evolutionists, both atheistic and theistic, who insist that it is a metaphysical principle and all reality must be explained in terms of evolution.

## Archbishop Sheen on Evolutionists' Idea of God

*Great Catholic Books Newsletter*, Vol. II, Number 2, was devoted to a short review of Archbishop Sheen's apostolic work. The excerpt below shows how that great thinker connected evolution-cult thinking with atheism:

> The first of his thirty-six published volumes, therefore, was a carefully documented analysis of the godlessness of modern civilization, especially in the United States. "Modern philosophy," it began, "has seen the birth of a

new notion of God...It is God in evolution. God is not. He becomes. In the beginning was not the Lord, but in the beginning was the Movement. From this movement God is born by successive creations. As the world progresses, He progresses; as the world acquires perfection, He acquires perfection. (Moreover) man is a necessary step in the evolution of God. Just as man came from the beast, God will come from man...It is the purpose of this work to examine this new notion of God," which he did, through 300 pages of quotations from American and English pragmatists like James, atheists like Dewey, naturalists like Sellars, and agnostics like Hume, Huxley and Hocking.

**Who is Francisco Ayala?**

At the Vatican-Templeton-Jesuit conference, Francisco Ayala, an evolutionary biologist, former president of the American Association for the Advancement of Science and longtime advisor to the Templeton Foundation, said intelligent design and creationism were "blasphemous" to both Christians and scientists. The embrace by some Church leaders of the ideas of scientists such as Dr. Francisco Ayala is a symptom of why the Church can't hold younger Catholics who are voting with their feet. Ayala, a Spaniard educated there, is a former Dominican who quit the priesthood in 1961, the same year he was ordained. If he really ever had the Faith he may have lost it in the seminary. He immediately left Spain to study at Columbia University under the famous evolutionist Theodosius Dobzhansky. Ayala is divorced and remarried. He served on the advisory board of the Campaign to Defend the Constitution, an organization that has lobbied Congress to lift federal restrictions on funding research on human embryos. In 2007 he was awarded the first of 100 bicentennial medals at Mount Saint Mary's University in Emmitsburg, MD for lecturing there as the first presenter for the Bicentennial

Distinguished Lecture Series. His lecture was entitled "The Biological Foundations of Morality." Dr. Ayala could be the 'poster child" for Catholic evolutionists. He has stated that creationism and intelligent design are not only pseudoscience, but also misunderstood from a theological point of view. He suggested that the theory of evolution resolves the problem of evil, thus being a kind of theodicy. A past recipient of the big bucks that come with the annual award of the Templeton Prize, he has said that his science "is compatible with religious faith in a personal, omnipotent and benevolent God." That statement reflects another reason people embrace evolution as the origin of humans, namely, because it provides moral autonomy that enables one to create God into whatever they want Him to be, such as a senile grandfather who blesses whatever ideas and behavior they chose.

### Smug Spanish Theologian

While on the subject of wayward Spaniards in the mainstream of Catholic theistic evolution, I refer back to the preface of her book, reprinted above, in which my wife mentioned that I had become annoyed after picking up a book by a theologian teaching at a Catholic university who was promoting evolution. It was a new book being sold in 2009 at a Catholic bookstore in Washington. The author, a priest at the University of Navarre in Spain, wrote to the effect that the only people who don't believe in evolution are "American Fundamentalists." Perhaps that priest reflects mainstream theology in Spain. Spain's popularly-elected Congress in 2005 made Spain the third country in the world to legalize homosexual marriage. It figures. Accept evolution, deny the Creator and that follows. (*Romans* 1:26-27). An article in *Forbes*, May 30, 2012, showed that Spain's economic disaster traced to its loss of Catholic family values and the resultant 50% drop in the fertility rate to one of the lowest in the world. (The collapse of the birth rate throughout contracepting-aborting

Europe has also opened the door to the Muslim takeover.) Those Americans disparaged in some Catholic intellectual circles as "Fundamentalists" are now the heirs of the Christian scientific intellectualism that was once the possession of Catholics and are the majority of those not bamboozled by Humanist evolutionary propaganda. Some Catholic homilists have been known to point out that the Evangelicals wouldn't have the Bible were it not for the Catholic Church. True enough; but I wonder if smugness on the part of some Catholic homilists is what keeps them from appreciating the efforts of Evangelicals to use natural science to restore faith in the Bible and refute the Humanists attack on it.

**The American Catholic Awakening**

Evangelicals are no longer the sole possessors of scientific intellectualism. In America, there is the growing stature of the Kolbe Center for the Study of Creation, founded in 2000 by the convert son of an atheist who was a former Secretary General of International Planned Parenthood. The Kolbe Center, "operating on a shoestring," so to speak, coordinates with an international group of Catholic scholars and has presented seminars in North America, Europe, Africa and Oceania. On the Center's website, kolbecenter.org, in addition to science and theology, there is the explanation of why St. Maximilian Kolbe is a patron. Many Catholics have heard how the saint offered his life to save another man's life in a Nazi concentration camp during WW II. What is lesser known is that he was a world-wide promoter of Our Lady under her title The Immaculate Conception. He founded in Poland what became the largest religious community in the world, with over 900 friars. He founded religious houses in Japan and India. He published a magazine in several languages, *The Immaculata*, with a circulation of over a million copies. And he was a severe critic of evolution:

> I cannot believe that man is only a perfect monkey. This is the question of evolution . . . A mountain of acute

criticisms has been published on this subject; but the more books they write the more complicated the problems grow. This theory not only does not agree with the results of today's experimental science, which is in constant progress, but in reality it contradicts these findings, as has been carefully documented.

## British and Irish Awakening

For Catholics in Great Britain and Ireland there is The Daylight Origins Society. http://www.daylightorigins.com Founded in Great Britain, the Society has been growing in recent years because of the additional support of inspired and active contributors in Ireland. In Appendix II there is a letter I received from one of those active contributors in Ireland in which he describes how he was a cradle Catholic who later lost faith in God's existence but had it restored because of American Evangelicals who went to Ireland and taught creation science.

Many have wondered how Ireland, that once Catholic country, so rapidly apostatized to the point wherein, in a May 2015 national referendum, 62% voted to change the Constitution to allow homosexual couples to marry. Ireland was the first country in the world to legalize same-sex marriage through a popular vote. In 2017 a homosexual whose father was from India became Prime Minister. Earlier in this chapter readers may recall they read of the incident this writer's wife described in the preface to her book in which she complained to the Headmaster Brother Kennedy that the school textbooks were teaching evolution as a fact. That incident was at a Christian Brothers school in Ireland in 1979. Evolution has been the doctrine taught in all of the life and physical science classes in all Irish Catholic schools since before 1979.

**Irish Youth Do Not Believe in God;**

It is a tragic irony of history that Irish Catholics resisted the British frontal attack on of their Faith for centuries but were so easily seduced by British and American Humanist education in less than three generations. When belief in evolution entered the doors of all of Ireland's schools and seminaries, the Faith went out the window. On the day after the pro-homosexual vote, Diarmuid Martin, the Archbishop of Dublin, said the Church in Ireland needed to reconnect with young people. The Archbishop told the Irish broadcaster RTÉ: "We [the Church] have to stop and have a reality check, not move into denial of the realities... I ask myself, most of these young people who voted 'yes' are products of our Catholic school system for 12 years. I'm saying there's a big challenge there to see how we get across the message of the Church." In Ireland most public schools are nominally Catholic schools; private schools are rare. Schools teach the doctrines of Humanism and that is not going to change. A 2013 survey of 1,146 college students by the Student Marketing Network found that only 37.5% believed in God and 83.5% believed abortion should be legal in Ireland. Read the survey report here: thejournal.ie/students-religion-ireland-1035328-Aug2013/. Why was the Archbishop surprised? See also the Oct.25, 2017 research report "The Faith Crisis of Today's Irish Youth" https://www.barna.com/research/faith-crisis-todays-irish-youth/

*LifeSiteNews* reported in July 2016 that Archbishop Martin, in response to the murder of a French priest by Muslims, was advocating tolerance and respect for other religions, particularly Islam. He said "Long-term solutions will come from education." Long term, with a little more education, Catholicism can be expunged in Ireland.

On the Sunday after the pro-homosexual vote in Ireland a U.S. pastor expressed shock that the "most Catholic people contradicted themselves." Why was he surprised about Irish Catholics? As is shown in the next chapter, according to a 2014 Pew Research survey, 57% of American Catholics favor homosexual marriage. That is just less than the 60% of mainline Protestants who do. Support among Catholics grew from 40% in 2007. The younger the population surveyed the more supportive it is. Catholic youth are included in the 67% of the "Millennials" born in 1981 or later who support it. The pastor blamed the "turncoat Catholics" of the U.S. Supreme Court for how they voted to legalize homosexual marriage and singled out Justice Kennedy whose house he said he had blessed and been his dinner guest. In all honesty American clergy should not suppose that young Irish Catholics did what young American Catholics wouldn't do if they had the choice. As explained in chapters 1 and 2 of this book, the Church in America is losing its youth in droves. Is it time for a U.S. reality check?

### Pew Research Findings on Homosexual Marriage

In Chapter 13 one will read that Cardinal Ratzinger in 1989 said that issues of sexual morality were due to a "false vision of humanity." The future Pope attributed that false vision to belief in evolution and the disappearance of the Church's doctrine on creation from theological education. Priests and parents are urged to seriously consider the impact of belief in evolution on our young. Could there be a connection between the Pew finding that American Evangelicals are only half as likely to believe in evolution and only half as likely to support homosexual marriage as American Catholics are? What will it take to wake up the Catholic Church in America to the seduction of its children by the unopposed indoctrination from the Humanist education industry through bogus evolution "science"?

# Chapter 12-Who made God?

This question is a major objection that Humanists put forward to justify their disbelief. Bertrand Russell (1872–1970), a famous British philosopher, in his influential little essay, *Why I am not a Christian,* put this forward as his first objection. Today's Humanists repeat the objection. For example, Philip Adams, famous in Australia as a humanist, social commentator, broadcaster, and public intellectual, spoke at the 2010 Global Atheists' Congress in Melbourne Australia:

> The great argument for God was that there had to be a Creation, a beginning. … But my objection was simple: If God was the beginning, who began God?

Humanists believe that the universe and life began with no adequate cause, contradicting rationality, because everything that has a beginning must have a sufficient cause.

This writer once passed along one of Stephen Jay Gould's preposterous statements to a hi-tech Ph.D. of his acquaintance who, when he read it, replied, "Yes, Gould is certifiably nuts." One wishes it was that simple, but it is far deeper. In 2001, Gould wrote an autobiographical essay for his last book from which one learns that he was the grandson of non-observant Hungarian Jews who arrived in New York in 1901. He began the essay, "I Have Landed," as follows:

> As a young child, thinking as big as big can be and getting nowhere for the effort, I would often lie awake at night, pondering the mysteries of infinity and eternity—and feeling pure awe (in an inchoate [imperfectly formed or developed], but intense, boyish way) at my utter inability to comprehend. How could time begin? For even if a God created matter at a definite moment, then who made God? An eternity of spirit seemed just as

> incomprehensible as a temporal sequence of matter with no beginning. And how could space end? For even if an intrepid group of astronauts encountered a brick wall at the end of the universe, what lay beyond the wall? An infinity of wall seemed just as inconceivable as a never-ending extension of stars and galaxies.

As the essay went on, he affirmed the faith choice he had made, namely, to believe in a "temporal sequence of matter with no beginning." With no adult to guide him to faith in God, and his school teachers' encouragement regarding evolution as the answer of science to his boyish questions, he developed non-theistic faith. Atheist superstar author, lecturer and evolutionary biologist Richard Dawkins asked the same question as Gould. In response to proponents of intelligent design he was reported to have asked: "Who designed the designer?" Who among the most faithful Catholics has not wondered "How did God get there?" Whatever can be known about God through reason and Revelation, it always requires a decision to say "I believe."

### God Never Hid

Still, God has never hidden Himself. In the Epistle to the Romans, St. Paul points out that the existence of God can be known by observing nature, i.e., through the things that He has made. That is a constant teaching of the Church and many are familiar with it. But, far fewer are familiar with the context in which that teaching lies. It is not one that makes it to the Sunday readings. It is an important part of Revelation because it explains why otherwise intelligent and well-educated people take positions and say things which cause educated Catholics to question their apparent lack of common sense. It also provides a plausible explanation for why homosexual activity and other vices are flaunted by homosexuals and approved of by the majority of non-homosexuals in our culture where the majority of the population denies its Creator. Verses 18-32 of that Epistle are:

> [18] For the wrath of God is revealed from heaven against all ungodliness and wickedness of those who by their wickedness suppress the truth. [19] For what can be known about God is plain to them, because God has shown it to them. [20] Ever since the creation of the world his eternal power and divine nature, invisible though they are, have been understood and seen through the things he has made. So they are without excuse; [21] for though they knew God, they did not honor him as God or give thanks to him, but they became futile in their thinking and their senseless minds were darkened. [22] Claiming to be wise, they became fools; [23] and they exchanged the glory of the immortal God for images resembling a mortal human being or birds or four-footed animals or reptiles.
>
> [24] Therefore God gave them up in the lusts of their hearts to impurity, to the degrading of their bodies among themselves, [25] because they exchanged the truth about God for a lie and worshiped and served the creature rather than the Creator, who is blessed forever! Amen.
>
> [26] For this reason God gave them up to degrading passions. Their women exchanged natural intercourse for unnatural, [27] and in the same way also the men, giving up natural intercourse with women, were consumed with passion for one another. Men committed shameless acts with men and received in their own persons the due penalty for their error.
>
> [28] And since they did not see fit to acknowledge God, God gave them up to a debased mind and to things that should not be done. [29] They were filled with every kind of

wickedness, evil, covetousness, malice. Full of envy, murder, strife, deceit, craftiness, they are gossips, [30] slanderers, God-haters, insolent, haughty, boastful, inventors of evil, rebellious toward parents, [31] foolish, faithless, heartless, ruthless. [32] They know God's decree that those who practice such things deserve to die—yet they not only do them but even applaud others who practice them.

Does anything in the above remind the reader of the state of American culture? Has the reader ever heard a priest preach on that text? Has that text ever made it to the First or Second Reading at Mass? Most of the rest of the epistle to the Romans has. Who censored verses 18-32? And why? Credible reports and criminal evidence suggests homosexual influences from the Vatican down to Episcopal chanceries worldwide.

## Catholics and Homosexualism

American Church officials and their diocesan media reacted as if the U.S. Supreme Court's June 2015 decision on homosexual "marriage" was a shock. Social research indicated that prior to that decision self-identified Catholics were as likely to favor, or at least to have no problem with, homosexual marriage as any other segment of the population. According to a September 24, 2014, report by the Pew Research Center, "Changing Attitudes on Gay Marriage,"

> In Pew Research polling in 2001, Americans opposed same-sex marriage by a 57% to 35% margin. Since then, support for same-sex marriage has steadily grown. Today, a majority of Americans (52%) support same-sex marriage, compared with 40% who oppose it...And among Catholics and white mainline Protestants, roughly six-in-ten now express support for same-sex marriage. Support for same-sex marriage also has grown among

> black Protestants. Support among white evangelical Protestants remains lower than among other religious groups.

Graphs included with the report provided additional support for the text. For example, 57% of Catholics favor homosexual marriage and that is just less than the 60% of mainline Protestants who do. Black Protestant approval is 41% and white Evangelical approval is 21%. Support among Catholics grew from 40% in 2007. The graphs also show that the younger the population surveyed the more supportive it is. Is there any reason to assert that Catholic youth are not representative of the 67% of the "Millennials" born in 1981 or later who support it? Graphs show that by a slight majority women support homosexual marriage while by a slight majority men oppose it.

Another Pew Research survey published March 15, 2015 asked if wedding-related businesses should be allowed to refuse those services to same-sex couples for religious reasons. Forty-seven percent said they should be allowed, forty-nine percent said they should not be allowed. Older Americans were heavily in favor of the right to refuse while 62% of those 30 and under said "no" to religious freedom. Just 40% of Catholics supported the right to refuse.

The Supreme Court's June 2015 decision was entirely predictable because in October 2014, the Supreme Court, by refusing to review it, affirmed the decision by a lower Federal court that struck down the ban on homosexual marriage that had recently been added to the Virginia Constitution by the vote of the citizens of Virginia. Upon receipt of that news in October 2014, the two bishops of Virginia issued a joint statement that expressed their "disappointment" and said

> It is our fervent hope that the Supreme Court will reconsider this fundamental issue in the future.

The Supreme Court did reconsider and made it final. One doesn't want to judge our bishops' efforts unkindly, but placing fervent hope in the Supreme Court to solve Catholic America's ambivalence about moral perversion seemed to lay observers to be misplaced hope. What if, instead of putting "fervent hope" in the secular courts of men to uphold natural law the appointed teachers had put their faith and hope in Divine Revelation? What if those 68% of white, non-Hispanic Catholics who believe that the fabulous natural world around us evolved over billions of years understood instead that it was instantly created out of nothing for us by the God who loves us without limit and that "scientific" theories to the contrary are bogus? Possibly, then, the majority of American Catholics would be giving glory to their Creator as St. Paul advised instead of to the animal ancestors from whom they believe they evolved.

If one accepts that humans evolved from animals through a natural process during millions of years, even with vague ideas of God "pulling the strings," how can he have anything but a false vision of what it means to be human? From the earliest days of civilization, humans have considered themselves exceptional among living creatures. But a 2016 survey by the Discovery Institute of more than 3,400 American adults indicates that the theory of evolution is beginning to erode that belief in humanity's unique status and dignity. According to the survey, 43 percent of Americans now agree that "Evolution shows that no living thing is more important than any other," and 45 percent of Americans believe that "Evolution shows that human beings are not fundamentally different from other animals." The highest levels of support for the idea that evolution shows that humans aren't fundamentally different from other animals are found among self-identified atheists (69 percent), agnostics (60 percent), and 18 to 29 year-olds (51 percent). Read the Survey here: evolutionnews.org/2016/04/new_poll_reveal102751.html

# Chapter 13-Diagnosis and Cure

The social research by Pew and others proves that the U.S. is imitating Europe in the rush to apostasy. Consider the diagnosis and a recommended cure that future Pope Benedict XVI made in a May 1989 address to the Presidents of the European Doctrinal Commissions. Speaking then as Prefect of the Congregation for the Doctrine of the Faith on the subject of "Difficulties Confronting the Faith in Europe Today," Cardinal Ratzinger traced through the litany of issues pertaining to sexual morality and the Church's sacramental order and said they are linked together by the <u>same false vision of humanity.</u> He went on to say that:

> We can give a proper answer to the conflict in detail only if we keep all of the relationships in view. It is their disappearance which has robbed the Faith of its reasonableness. In this context, I would like to list three areas within the worldview of the Faith which have witnessed a certain kind of reduction in the last centuries, a reduction which has been gradually preparing the way for another "paradigm."
>
> **In the first place, we have to point out the almost complete disappearance of the doctrine on creation from theology**.[Emphasis added.] As typical instances, we may cite two compendia of modern theology in which the doctrine on creation is eliminated as part of the content of the faith and is replaced by vague considerations from existential philosophy, [he then named two published in Europe]. In a time when we are experiencing the agonizing of creation against man's work and when the question of the limits and standards of creation upon our activity has become the central problem

> of our ethical responsibility, this fact must appear quite strange. Notwithstanding all this, it remains always a disagreeable fact that 'nature' should be viewed as a moral issue...That nature has a mathematically intelligibility is to state the obvious, the assertion that it also contains in itself a moral intelligibility, however is rejected as a metaphysical fantasy. The demise of metaphysics goes hand in hand with the displacement of the teaching on creation.

In *Aquinas and Evolution* by Fr. Chaberek (introduced in chapter 8) there is a paragraph that may explain what Ratzinger meant when he referred to "a reduction which has been gradually preparing the way for another 'paradigm'."

> As we noted, it is not the understanding of Aquinas or evolution that has changed over the last century or so. It is rather the change in paradigms—from roughly speaking 'Biblical' or 'creationist' to 'naturalistic' or 'evolutionary'. This change of paradigms explains why a great number of today's Thomists greatly differ from those of a century ago. In our opinion, the 'evolutionary' as opposed to the 'Biblical' is not the proper context in which the problem of origins should be addressed. For this reason we believe that not today's, but the previous Thomists were closer to the truth regarding both—the interpretation of Aquinas's metaphysics and the assessment of the evolutionary theory of origins...
>
> In what follows we will show that the teachings of Thomas Aquinas—and indeed any sound philosophy...are not just incompatible with the Darwinian theory but exclude it in principle. By showing this we want to achieve another objective, namely, to help contemporary Thomists to realize some of the difficulties, inaccuracies, or even flat-out errors in their interpretation

of Aquinas when it comes to the origin of species and man. (Page 12.)

**What Priests Can Preach Creation?**

Six years after Cardinal Ratzinger's 1989 speech to the European bishops, when he published a collection of homilies he had given under the title *In the Beginning...': A Catholic Understanding of the Story of Creation and the Fall*, he wrote in its preface that "...the creation account is noticeably and completely absent from catechesis, preaching, and even theology. The creation narratives go unmentioned; it is asking too much to expect anyone to speak of them."

Are any theologians, parish priests, bishops, Catholic school officials or rectors of seminaries listening? Are those who volunteered and allowed themselves to be appointed as our spiritual shepherds going to do their job in this regard?

Beside the misinterpretation of Aquinas, at least one other reason priests don't preach about our creation doctrines is that the historicity of *Genesis* has been taught to them in the seminary as goofy symbolism as I illustrated by the analysis in chapter nine.

A truism often repeated in homilies, stated in Church documents, and popular Catholic writing is that there is no conflict between scientific truth and religious truth because God is the author of both. Yet, as Cardinal Ratzinger lamented, the religious truth of creation goes unmentioned and has almost completely disappeared from theology books, catechesis, and preaching. Why so? Can our Catholic intellectuals, lay and clerical, give a reason other than their faith in evolution contrary to *Humani Generis* and natural science? Is it that they, like the ordinary Catholics, have been misled by the "blind guides" before them? Is it fear of speaking out and being ridiculed as a "backward creationist"? The truth is well within their reach. The evolutionists' only tactic is to abuse their opponents because the facts are not on their side. One example of the Humanist diversionary tactics is to ask "What about Galileo?"

**Humanists Intimidate Via the Galileo Myth**

In Chapter Eight the reader found the text of *Humani Generis* unambiguously presented. Some readers may still be half afraid to believe that the Church could be so right and the Humanist scientific consensus so wrong because of things such as the popular narrative regarding the 17th Century "Galileo case" that is imbedded in our secular culture and propagated by scientific method-Catholic scholars. In his Foreword to Fr. Chaberek's book, philosophy professor Logan Gage called it the "Galileo Complex." He wrote that "this neurosis might be cured by a better understanding of the Galileo affair." The Galileo incident has been extrapolated through the years to the effect that the Catholic Church demonstrated then and there that it has no competence whatsoever to even speak of science. According to the popular narrative, the Church through the "dreaded Inquisition," condemned Galileo and mistreated him for his avocation of well-known and proven scientific facts that clashed with Catholic interpretation of Scripture. That false story with various embellishments has circulated for nearly 400 years. Numerous books have been written about it and references are made to it in literature and theatre. In Chapter nine I explained how heterodox Fr. Guinan said "the traditional views of Genesis have suffered three challenges: Galileo on the movement of the earth around the sun and not vice versa; the growth of geology in the 18-19th centuries [Hutton, Lyell] and discoveries [hypothesizes really] about the age of the earth; and Darwin's theory of evolution." I've dealt with all of those bogus "challenges" so far except for the false Galileo narrative promoted by Humanists and their fellow travelers. I will deal with that as explained at the end of this chapter but first get to know another Jesuit who promulgates the Galileo Complex.

**Jesuit Cosmologist**

Another Catholic who contributes to Humanist propaganda via the false Galileo story is papal astronomer Brother Guy Consolmagno, S.J. who is in the mainstream of Jesuit-papal astronomers such as George V. Coyne, S.J. who is based at the U. of Arizona where Br. Consolmagno earned a doctorate in Planetary Science in 1978. (Fr. Coyne was discussed in Chapter 5.) Dr. Consolmagno became a Jesuit in 1989 and became a Vatican Observatory astronomer in 1993. His research explores connections between meteorites, asteroids, and the evolution of small solar system bodies. He has impeccable credentials within the evolutionary cosmology club. For example, he was chair of the American Astronomical Society's Division for Planetary Sciences. In 2014, he was the first clerical awarded the Carl Sagan Medal for public outreach by the American Astronomical Society Division for Planetary Sciences. One can't ascribe to Br. Consolmagno the Humanist beliefs of Carl Sagan but it is difficult to disassociate a scientist from the Society in which he held a leadership position. It would be wrong to make any definite assertions about the Vatican's astronomer's beliefs except those that are revealed in his published works. From Now You Know Media one can buy Brother Consolmagno's video DVD set "Galileo: Science, Faith, and the Catholic Church." According to part of the promotional material for the boxed set:

> The Galileo affair resonates with our own times. Although the debate about an earth or sun-centered universe is long past, the ways we react to new ideas hasn't changed at all. All of the hopes, fears, and misunderstandings that surrounded Galileo and his opponents, we still face today in our encounters with science and religion. By spending time with Galileo and

his story, you will enrich your own faith and increase your understanding of science and religion.

Surely Jesuit Consolmagno was not including himself in the "we" whose way of reacting "to new ideas hasn't changed at all" since the 17th Century. What is that statement but a not very subtle criticism of the Magisterium and the rest of us for not adopting the 18th Century "science" of Immanuel Kant about how the sun and stars formed themselves out of eternally-existing "clouds of nebular gas" and goofy "Big Bang" Theories that he and the evolutionary cosmologists of the American Astronomical Society Division for Planetary Sciences peddle for our belief?

**The True Story of Galileo**

The Galileo Case resurfaced most dramatically and was splashed throughout the world by the media when in 1992 Pope John Paul II issued what was interpreted as a formal apology on behalf of the Catholic Church for its scientific incompetence in the Galileo affair. Nothing could be further from the truth. The Vatican Congregation involved judged and acted entirely reasonably based on the facts of 1616. In fact, science wasn't even the primary issue of the Case. To explain the details of the Galileo affair and Pope John Paul's 1992 "apology" required so many pages that I have put it online to reduce the pages of this book and thus make this book more affordable. In Appendix I the true story of the Galileo Case is introduced and a reference is given to where the full story can be downloaded from the web for reading, copying or printing. Anyone who thinks the Galileo affair in any way reflected badly on the Church or has ever been intimidated by it should read it.

In the next chapter, the means by which unproved and implausible theories of origins contrary to Holy Scripture became the accepted gospel of Humanists and the majority of Catholic adults in the United States will be explained.

# Chapter 14-Humanism Crosses the Atlantic

"*As the culture war is about irreconcilable beliefs about God and man, right and wrong, good and evil, and is at root a religious war, it will be with us so long as men are free to act on their beliefs.*"- Patrick Buchanan

The purpose of this chapter is to answer this question: "If the theory of evolution is just so much scientific bunk as you have said, how can you explain that it has become accepted as the scientific consensus and sixty-five percent of American adults believe it?" The short answer to that question is that people believe what they have been taught by the American education institutions such as schools and universities and are also affected by cultural influences. The long answer requires an explanation of how Humanist philosophy has dominated America since its founding and how Humanists have both converted and neutralized Catholics. Dennis Q. McInerny, when he was Professor of Philosophy at Our Lady of Guadalupe Seminary, summarized that conversion process as follows:

> Over the course of the past century and a half, Western society has allowed itself to be convinced by something which, from a strictly scientific point of view, is singularly unconvincing. I speak of the theory of evolution. But if this theory fails to make the grade as serious science, it has managed to succeed spectacularly as a philosophy, a comprehensive worldview, whose presence is pervasive and whose influence is as powerful as it is deleterious. Its invasion of our educational system is complete, and for decades now the nation's youth have been systematically indoctrinated to accept as an unquestionable "fact" what, in fact, is anything but.

This writer hopes he has demonstrated in this book that Humanism is a non-theistic religion based on faith in an uncreated, self-existing universe, and the biological macroevolution of humans without immortal souls. Those principles are laid out in *Humanist Manifesto I.* Humanist philosophy is a particular set of ideas about knowledge, truth, and the nature and meaning of life, based on Humanist faith. It is the Humanist worldview based on explanations of reality that exclude God and substitute unobserved and implausible mechanisms they call "science." Catholicism is a faith based on logical considerations, such as cause and effect and that is informed by Divine Revelation. Philosophy grounded on Catholicism is a particular set of ideas about knowledge, truth, and the nature and meaning of life. There is a Catholic worldview.

These faiths cannot co-exist. Humanists won't tolerate it. Moral evil is institutionalized in America under the force of civil law. Pat Buchanan explained succinctly that co-existence is not the goal of Humanists. In an April 2015 column titled "The Long Retreat in the Culture War." Buchanan observed that "Christianity, driven out of schools and the public square, is being whipped back into the churches and told to stay there." He recounted how the Humanists start by appealing for tolerance.

> First comes a call for tolerance for those who believe and behave differently. Then comes a plea for acceptance. Next comes a demand for codifying in law a right to engage in actions formerly regarded as debased or criminal. Finally comes a demand to punish any and all who persist in their public conduct or their private business in defying the new moral order.

And so it has gone. Buchanan observed that "a Christian majority that had the Faith that created Western civilization behind it rolled over and played dead. Christians watched paralyzed as their country was taken from them." If only Christians had the passion that the Humanists have shown in their religious rage at Trump's reform efforts.

**There is No Salvation in Politics**

Humanism is a religion that imposes a political regime in the same way that Islam does. Therefore, Christians must combat it as a false religion rather than through secular politics based on one or more single issues as they have been doing at least since *Roe v Wade*. That abortion, euthanasia and all of the aspects of the "Culture of Death" are the result of Darwinism was well-demonstrated in *From Darwin to Hitler: Evolutionary Ethics, Eugenics, and Racism in Germany,* a 2004 book by Richard Weikart, Associate Professor of History at California State University, Stanislaus, CA. He gave a lecture at U. Cal Santa Barbara called "From Darwin to Hitler" that can be viewed on YouTube in which he discussed six "Implications of Darwinism for Devaluing Human Life."
youtube.com/watch?v=w_5EwYpLD6A

It is 58 minutes and a real "eye opener." Professor Weikart noted that "the devaluing of human life will continue as long as Darwinism is ascendant." Those who have worked in the pro-life movement must realize that the other side can't be reached by argument. Those who deny the sanctity of life, on the basis of 'stark raving rationalism', have undergone a conversion experience that shuts off any return to pre-conversion sensitivity to sanctity of life arguments, even arguments of a prudential kind. Those who are married to the pro-abortion view would rather be "biting into their wrists" than give up that belief. Professor Weikart's 2016 book, *The Death of Humanity: and the Case for Life* explores our culture's declining respect for the sanctity of human life, drawing on philosophy and history to reveal the dark road ahead for society if we lose our faith in human life. Humanists have clearly expressed their goals and have seen most of those become the public policy and culture. Mark Steyn observed in *After America* that

> The United States has not just a ruling class, but a ruling monoculture. Its 'truth' and 'facts' and 'science' permeate not just government but culture, the media, the institutions in which we educate our children, the language of public discourse, the very societal air we breathe. That's the problem, and just pulling the lever for a guy with an R after his name every other November isn't going to fix it.

Steyn explained that "changing the culture (the schools, the churches, the movies, the TV shows) is more important than changing the politics." The election of Trump will not save us.

**You Can't Beat Them by Accepting Their False Doctrine**

A false religion can't be defeated as long as Catholics continue to accept its fundamental doctrine. As has been shown in earlier chapters, especially Chapters 8 and 13, Catholic philosophy has been polluted by acceptance of in varying degrees the basis of the Humanist faith, namely, cosmic and biological macroevolution. It will be shown in this chapter how Humanists ideas developed and how Humanists worked to have their philosophy become the public policy of the United States. In this chapter, the writer traces the history of Humanism and illustrates the harmony between Humanist philosophy and present U.S. public policy.

**America's Most Influential Philosopher and Educator**

The central character in this history is America's most influential philosopher and educator, John Dewey. The people and ideas that influenced Dewey will be related, and his subsequent influence in making America's policy Humanist will be told. If the reader is not familiar with John Dewey, or perhaps never heard of him, the opening paragraphs of a long article, "John Dewey: Prophet of American Naturalism," by Fr. John Hardon, published in the September 1952 issue of *The Catholic Educational Review* will serve as an introduction:

> When John Dewey celebrated his ninetieth birthday on October 20, 1949, fifteen hundred guests crowded a huge ballroom in New York City to do him honor. Messages of congratulation poured in from President Harry Truman, Prime Minister Atlee, Pandit Nehru, and from a hundred United States colleges and universities. A dozen foreign nations had planned celebrations. Friends were raising $90,000 [a lot of money in 1949] for an educational Dewey Birthday Fund. And all because in the eyes of millions of admirers no one in the history of America has so profoundly and in so many areas of human endeavor influenced and determined his own age as . . . America's dean of Philosophers: John Dewey.
>
> In striking contrast with this adulation, American Catholics regard Dewey as a modern prophet of error whose philosophy of education is socialistic naturalism without God, without Christ, without religion, without immortality. Every single strain in it, from the influence of Hegel to the inspiration of Darwin, finds its place within his system.

**Dewey's Story is Representative**

In telling Dewey's story, this writer hopes to communicate that it was not just Dewey personally, but also his disciples and like-minded Humanists teaching at the most influential universities in America, who spread Humanism and enforce its code. Effectively, many universities function as "temples" of the Humanist faith. While Dewey is a central character, his education, attitudes, convictions, and non-theistic faith were similar to those of the 19th Century and early 20th century intellectuals who shaped American education. It would almost be impossible for evolution to not be accepted by most of the intellectuals of the last 100 years. As will be seen, acceptance of evolution was not based primarily on an examination of its scientific claims, but on the way it harmonized with philosophy developed by men entirely without access to sanctifying grace, knowledge of God, and the gift of faith in Him.

## The Roots of Humanist Philosophy

The trail of Humanism from Europe to the colonial and early United States runs through Great Britain. As will be seen, the U.S. East Coast, particularly New England, was fertile ground for Humanism because its chaotic religious culture and many of its institutions were similar to those in "the mother country." An understanding of that "crazy quilt" religious history will explain why so many of the university-educated of the 19th Century lost all belief in God and essentially reverted to the beliefs of the ancient Greeks. It is no coincidence that New England has been the most pro-abortion region of the U.S. A Pew Research survey in 2013 found that 75% of New Englanders believe abortion should be legal all or most of the time.

Not just a few Catholic scholar-priests, such as Fr. Bruce Vawter and Fr. Michael Guinan, have spent their whole adult lives practicing the "scientific method" to determine how those fantastic stories in the Bible originated. It might be worthwhile to consider how those even more fantastic stories in the "Evolutionist gospel" preached by Darwin's disciples such as Stephen Jay Gould and Stephen Hawking originated.

## What Pagan Greeks Believed

Humanist philosophy harkens back to the Greek philosophers. Many of those believed in evolution and an ancient Earth. In fact it is almost astonishing to realize that in the 21st century, the philosophy-driven scientific consensus has so much in common with them. The early 20th-century evolutionist Henry Fairfield Osborn, a university science professor and for 25 years director of the American Museum of Natural History, showed in his book, *From the Greeks to Darwin* (NY: Charles Scribner's Sons, 1929) that all of the essential ideas of Darwin's theory can be found in the writings of the ancient Greeks. Founded on the discovery of Piltdown Man, Osborn developed his own evolution theory of man's origins called the "Dawn Man Theory." Writing before Piltdown was exposed as a hoax, Osborn maintained that *Eanthropus* or "Dawn Man" sprang from a common ancestor with the ape.

Anaximander (610-540 B.C) believed that the universe had begun as an undifferentiated mass, from which all things had arisen by the separation of opposites; that all of the planets had once been fluid but had been evaporated by the sun; that life began in the sea but had been driven upon the land by the subsistence of water; that of those stranded animals some had developed the capacity to breathe air; and so had become the progenitors of all later life. He believed that man could not have been what he now was, for if man, on his first appearance, had been so helpless at birth, and had required so long adolescence, as in these later days, he could not have survived. Empedocles, in Sicily (445 B.C.), thought that organs rise not by design but by selection. Nature makes many trials and experiments with organisms, combining organs variously. Where the combination meets environmental needs, the organism survives and perpetuates. Where the combination fails, the organism is weeded out so that as time goes by the organisms are more intricately and successfully adapted to their surroundings. If any of this sounds familiar, it should. It's not far from Darwinism.

Democritus (circa 400 B.C.) believed there have been or will be an infinite number of worlds, that at every moment planets are colliding and dying, and new worlds are rising out of chaos by selective aggregation of atoms of similar size and shape. He held that there is no design and the universe is a machine. If any of this sounds familiar it should. It's not far from Stephen Hawking's cosmology.

With the founding of the Catholic universities in Europe, the best of the Greeks were utilized and refined in the light of Divine Revelation by great minds such as Albert the Great, Bonaventure, and Peter Lombard. Thomas Aquinas' *Summa Theologica* is considered to be the best example of scholastic, medieval, and Catholic philosophy.

## The Humanist Revival

The breakup of European Christendom began in 1517. The trigger was failed priest Martin Luther in Germany. In the period following the start of the Protestant Reformation/Revolt in 1517, Humanism began to revive as the result of philosophers in Germany. When men are separated from the sanctifying grace of the sacraments, deprived of the truth of faith and morals, and taught that everything they need to know is in the Bible which they are free to interpret for themselves, bad things happen. Bad things normally start with bad ideas, and a series of German and British philosophers provided plenty of those. If the most influential of those were not all actual atheists, their ideas were so far removed from what began in 1517 that the original Protestant "reformers" and Henry VIII would have been shocked.

Elizabeth I reigned from 1558 to 1603 and decimated what was left of Catholicism in Great Britain. Britain became a religion cafeteria and fertile ground for atheism, agnosticism and deism. A paper, "Philosophical Naturalism and the age of the earth: are they related," gives an excellent history of the loss of faith in the

Bible and the substitution of philosophical naturalism by the leading men and Anglican clergy in 18th century Britain. (See http://creation.com/philosophical-naturalism-and-the-age-of-the-earth-are-they-related )

In addition to atheism, agnosticism and deism, Britain exported various religious sects to its North American colonies. In the 18th Century, there were a variety of sects espousing all sorts of beliefs, including some that were not even Christian. In the colonial period, the settlers were mainly British and Catholicism was banned.

Congregationalism was carried to America in 1620 by the Pilgrims who were members of John Robinson's congregation in Holland, originally of England. In New England, numerous communities were established based on Congregational-type religious principles. In 1648, in the Cambridge Platform, a summary of principles of church government and discipline was drawn up. But each local church remained free to make its own declaration of faith and free to decide its own form of worship; in the conduct of the local church each member was granted an equal voice. As the country expanded, Congregational churches were established in the newly opened frontier regions. Congregationalists were always prominent in education. They founded Harvard, Yale, Williams, Amherst, Oberlin, and many other colleges.

Universalists were well established in 18th Century America as an amalgamation of various European sects that believe that the God of love would not create a person knowing that that person would be destined for eternal damnation. They concluded that all people must be destined for salvation. At first they believed that people might “serve time” in hell, but later, hell was eliminated altogether. They claim their belief goes back to apostolic days, but a book by Gerrard Winstanley, *The Mysterie of God Concerning the Whole Creation, Mankinde* (London, 1648) "kick started" the modern movement. Universalists founded colleges as well, for example, Tufts University.

Unitarians started in Eastern Europe as a Christian heresy that denied that Jesus is God. They said He was a good moral teacher and a prophet of God. They eventually dropped most other Christian doctrines as the religion migrated through England to New England in 1782. From 1805, the Harvard Divinity School taught Unitarian theology. Many self-identified Unitarians were among the signers of *Humanist Manifesto I* in 1933.

### Christian Science, Mormonism

Christian Science is another religious cult that developed in late 19th Century New England. Mary Baker Eddy wrote a book called *Science and Health* in 1875. That book along with the King James Bible became the central text of the Christian Science sect. According to Eddy's book, which has sold over 9 million copies, sickness is an illusion that can be corrected by prayer alone. Christian Science theology is closer to Buddhism than Christianity, believing that reality is purely spiritual and the material world an illusion. Eddy's *Science and Health* reinterpreted key Christian concepts including the Trinity, the divinity of Jesus, atonement and resurrection. For example, the Holy Ghost is Christian Science, Jesus is a Christian Scientist "way-shower" between humanity and God, the crucifixion was not a divine sacrifice for the sins of humanity, there is no doctrine on the soul and heaven and hell are states of mind. Eddy viewed God not as a person but as "All-in-all." In May 1885 the Boston correspondent for the *London Times* wrote about the "Boston mind-cure craze": "Scores of the most valued Church members are joining the Christian Scientist branch of the metaphysical organization, and it has thus far been impossible to check the defection."

Mormonism is a non-Christian pantheistic religion founded by Joseph Smith in the 1820s in New York. It characterizes itself as the only true form of the Christian religion since the time of a Great Apostasy that began not long after the ascension of Jesus Christ. Most Americans probably think Mormons are Christians.

### Organized Non-Theistic Religions in America

Unitarians and Universalists merged into the Unitarian Universalist Association (UUA) in 1961. According to UUA's information on Wikipedia

> Unitarian Universalism is a liberal and syncretic religion characterized by a "free and responsible search for truth and meaning." The theology of individual Unitarian Universalists ranges widely, with the majority being Humanist but also having members that follow atheism, agnosticism, pantheism, deism, Judaism, Christianity, neopaganism, Hinduism, Buddhism, Taoism and many more.

Most Humanists are unlikely to be "card-carrying members" of an organized non-theistic religion, but some are. These religions provide an outlet for Humanists. For example, Stephen Jay Gould frequently spoke at a Unitarian Meeting House in Harvard Square.

Ethical Culture is another of the non-theistic organized religions. A precursor to the doctrines of the ethical movement can be found in the South Place Ethical Society, founded in a Unitarian chapel in London in 1793. The Ethical movement was another outgrowth of the general loss of faith among the intellectuals of the 19th Century. The foundation of Ethical Culture in America is credited to Felix Adler, the son of a prominent New York rabbi. As part of his education, he enrolled at a German university, where he was influenced by neo-Kantian philosophy. He was especially drawn to the Kantian ideas that one could not prove the existence or non-existence of deities or immortality and that morality could be established independently of theology. Upon his return from Germany, in 1873, he shared his ethical vision with his father's congregation in the form of a sermon. Due to the negative reaction he elicited, it became his first and last sermon as a rabbi in training. Instead he took up a professorship at Cornell University and, in 1876, gave a follow up sermon that led to the 1877 founding of the New York Society for Ethical Culture, which was the first of its kind. By 1886, similar societies had sprouted up in Philadelphia, Chicago, and St. Louis. These societies all adopted the same statement of principles including the belief that morality is independent of theology.

In effect, the movement aimed to "disentangle moral ideas from religious doctrines, metaphysical systems, and ethical theories, and to make them an independent force in personal life and social relations."

The movement does consider itself a religion in the sense that "Religion is that set of beliefs and/or institutions, behaviors and emotions which bind human beings to something beyond their individual selves and foster in its adherents a sense of humility and gratitude that, in turn, sets the tone of one's world-view and requires certain behavioral dispositions relative to that which transcends personal interests." A good illustration of non-theistic religious fervor is the "Global Warming" dogma.

Since around 1950, the Ethical Culture movement has been increasingly identified as part of the modern Humanist movement. Specifically, in 1952, the American Ethical Union, the national umbrella organization for Ethical Culture societies in the U.S., became one of the founding member organizations of the International Humanist and Ethical Union.

## Naturalism Meets Freemasonry

The brief outline above illustrates what happens when men are cut off from the grace of the Sacraments and the Church's Magisterium: multiplication of theistic and non-theistic sects characterized by belief that man can perfect himself and reform the social and political order in which they live. These ideas were most popular among the educated class who learned them in colleges in Great Britain and early America. Adding to the zoological garden of false religions that provided the fertile ground for the growth of Humanism in America was the importation of British Freemasonry. British Freemasonry was alive and spreading from the colonial days. With the disintegration of religious truth in Great Britain, many who became Naturalists or Humanists following Hutton, Lyell and others also became Freemasons. Freemasonry may be thought of as "Highly-Organized Militant Humanism" in the sense that it is a secretive organized hierarchical membership movement, holding doctrines and political aims that are indistinguishable from Humanism. Freemasonry was active in Scotland as early as 1598, in London in 1717, and The Premier Grand Lodge of England appointed a Provincial Grand Master for North America in 1731. Benjamin Franklin held that post in 1734. Many signers of the founding documents of the United States were Freemasons and 15 U.S. Presidents were known Freemasons.

## Condemned

Freemasonry was condemned by a papal encyclical in 1738 because it teaches a naturalistic religion, i.e., Humanism. Six more Popes wrote against it before 1884 when Pope Leo XIII wrote the encyclical *Humanum Genus* (On Freemasonry and Naturalism). In this context, Naturalism is evolution applied to philosophy. In reading what Pope Leo wrote, see if you notice anything he mentions happening in Humanist-dominated America today. The following are excerpts so for ease of reading I have omitted the ellipses and other punctuation that would interrupt the flow of the Pope's teaching.

> No longer making any secret of their purposes, they are now boldly rising up against God Himself. They are planning the destruction of holy Church publically and openly. The sect of Freemasons grew with a rapidity beyond conception in the course of a century and a half, until it came to be able, by means of fraud and audacity, to gain such entrance into every rank of the State as to seem to be almost its ruling power. Now, the fundamental doctrine of the naturalists is that human nature and human reason ought in all things to be mistress and guide. They allow no dogma of religion or truth which cannot be understood by the human intelligence. By a long and persevering labor, they endeavor to bring about this result--namely, that the teaching office of the Church and authority of the Church may become of no account in the civil State and for this same reason they declare to the people and contend that church and state ought to be altogether disjointed. By this means they reject from the laws and from the commonwealth the wholesome influence of the Catholic religion and they consequently imagine that States ought to be constituted without regard for any laws and precepts of the Church.
>
> Nor do they think it is enough to disregard the church unless they injure it by hostility. Indeed, with them it is lawful to attack with impunity the very foundations of the Catholic religion, in speech, in writing, and in teaching. And even the rights of the Church are not spared. The least possible liberty to manage affairs is left to the

Church and this is done by laws not apparently very hostile but in reality framed and fitted to hinder freedom of action.

Naturalists are carried to extremes, either by reason of the weakness of human nature, or because God inflicts upon them the just punishment of their pride. Hence it happens they no longer consider as certain and permanent those things which are fully understood by the natural light of reason, such as certainly are--the existence of God, the immaterial nature of the human soul, and its immortality. When this fundamental truth has been overturned and weakened, it follows that those truths, also which are known by the teaching of nature must begin to fall--namely, that all things were made by the free will of God the Creator, that the world is governed by Providence, that souls do not die, that to this life of men upon the earth there will succeed another and an everlasting life.

When these truths are done away with, which are as the principles of nature and important for knowledge and for practical use, it is easy to see what will become of both public and private morality. If these be taken away, as the Naturalists and Freemasons desire, there will immediately be no knowledge as to what constitutes justice and injustice, or upon what principle morality is founded. And, in truth, the teaching of morality which alone finds favor with the sect of the Freemasons, and in which they contend the youth should be instructed, is that which they call 'civil' and 'independent' and 'free,' namely, that which does not contain any religious belief. For, wherever this teaching has begun more completely to rule, there goodness and integrity of morals have begun quickly to perish, monstrous and shameful opinions have grown up, and the audacity of evil deeds has risen to a high degree. What refers to domestic life in the teaching of the Naturalists is almost all contained in the following declarations: that marriage belongs to the genus of commercial contracts, which can rightly be revoked by the will of those who made them, and that civil rulers of the State have power over the matrimonial bond; that in the education of youth nothing is to be taught in the matter of religion as certain and fixed opinion. To these things the Freemasons fully assent; and not only assent, but have long endeavored to make them into law and institution. Thus the time is quickly coming when marriage will be turned into another kind of contract---that is, into changeable and uncertain unions which fancy may join together, and which the same when changed may disunite.

What, therefore, the sect of the Freemasons is, and what course it pursues, appears sufficiently from the summary We have briefly given. Their chief dogmas are so greatly and manifestly at variance with reason that nothing can be more perverse. To wish to destroy the religion and the Church which God Himself has established and to bring back after a lapse of eighteen centuries the manners and customs of the pagans, is signal folly and audacious behavior.

**Freemason Nation**

Wow, Leo XIII predicted the "Freemason Nation" in which we now live. Can Christians co-exist? They won't let us. Like all secret societies, Freemasonry is swayed by men who only reveal the part of their intentions which they consider advisable to disclose. Their main goal is to recruit followers and the main method they apply is to gain control of the schools and universities in a country. This, of course, is the more easily denied as those who are in charge of applying the method are the first ones to deny it. In 1894, Pope Leo XIII followed up with an apostolic letter to all the governments of the world in which he attributed to Freemasonry the ambition of getting complete political control of each and every state, so as ultimately to become the supreme ruler of the world. On Wikipedia, one may find "List of Freemasons." Lists of Freemasons are in alphabetical order and include many famous names.

In summary, from its founding in colonial times, America's leading men have been the product of a culture formed by zany religions, naturalism, evolutionism, and Freemasonry. And it was into that culture that John Dewy, America's most influential philosopher, was born in 1859, the year Darwin published *The Origin of Species by Means of Natural Selection or the Preservation of the Favored Races in the Struggle for Life.*

**Dewey's Formation: Liberal Religion and British Periodicals**

John Dewey was born into a middle-class family in Burlington, VT. His early life in a small town in Vermont provided him an easy-going manner and politeness that made him very likable to colleagues and students. His mother was a Congregationalist. His biographer observed that "Dewey did not abandon all religion at that time in reaction to his mother's narrow pietism and excessive religious emotionalism" because of "the liberal evangelism he was finding in church and college" that was more palatable to his intellect. At sixteen, Dewey entered the University of Vermont in 1875. Founded in 1791, the university was Congregationalist.

The few science classes he took were "new and interesting, especially the biological sciences that touched upon the controversies of evolution current in academic and religious circles." He was a serious reader, and his favorite periodicals were three from England: *Fortnightly Review, Nineteenth Century,* and *Contemporary Review.* These published articles on economic, political, social, moral, religious, and philosophical problems by evolutionists and other Humanists such as, Charles Darwin; Thomas "Darwin's Bulldog" Huxley; Alfred Wallace who conceived the idea of evolution by natural selection and published it with Darwin in 1858; John Tyndall, a brilliant physicist who was a member of a club that vocally supported Darwin's theory of evolution and sought to strengthen the separation between religion and science; Leslie Stephen, an Anglican clergyman who renounced his religious beliefs and wrote *The Science of Ethics* (1882) which was extensively adopted as a textbook on the subject and made him the best-known proponent of evolutionary ethics in late-nineteenth-century Britain; George Lewes, a philosopher who became part of the ferment of ideas which encouraged discussion of Darwinism, positivism and religious skepticism; and Frederic Harrison, a famous proponent of positivist philosophy which holds that sensory experience is the exclusive source of all authoritative knowledge and rejects intuitive knowledge, metaphysics and theology.

**Influenced By Reading**

In these three periodicals there regularly appeared Humanist articles under such titles as "The Metaphysics of Materialism," "Modern Materialism: Its Attitude toward Theology," "An Agnostic's Apology," "The Skepticism of Believers," "The Religion of Positivism," "The Place of Conscience in Evolution," "Evolution as the Religion of the Future," and "The New Psychology." The theme of these magazines was trust in science and non-trust in religion. The underlying assumption was that these were the convictions of all intelligent men. It is easy to understand how a young man, barely beyond his teens with heretical Christianity upbringing and no sanctifying grace, became what he read.

In the same time period there was also rising an evolution-based genre of books, explaining to the unenlightened how religion evolved. One that comes to mind is Canadian novelist Grant Allen's 1897 book published in England, *Evolution of the Idea of God: an Inquiry into the Origins of Religion.* It's so blatantly fictional in its explanation of how Christianity "evolved" that it is almost funny to read. But people who didn't know any better took it seriously based on its pretention of scholarship.

**Rise of Humanist Philosophy in America**

This writer speculates that some Catholics who have had training in Thomism or other Scholastic Philosophy might think that Philosophy is a "Catholic thing." The reality is that the Philosophy Department of the majority of the world's universities and the majority of scholarly publishing about philosophy is Humanist. And it was so in the 19th Century when America's leading men were being formed. Dewey's "major' in college was political and social philosophy and his professor was a Congregationalist clergyman.

Upon graduation Dewey taught high school while submitting philosophy articles to Humanist academic journals. That got him noticed and led to a scholarship at Johns Hopkins University, where he majored in philosophy and obtained a doctorate. The president of the university discouraged Dewey from majoring in philosophy because at that time American colleges and universities only employed as philosophy teachers those trained in "Christian theology," whatever that was. There were so many Christian and quasi-Christian sects it could be anything.

At Hopkins in 1882, Dewey was mainly influenced by a guest philosophy professor from the University of Michigan who had studied in Germany and communicated the ideas of German philosophers Hegel and Kant as well as Spinoza, an agnostic Portuguese Jew based in the Netherlands. Dewey studied the pagan Greek philosophers, including Democritus, whose ideas about the universe as a guide to Stephen Hawking were discussed earlier in this chapter. Readers are encouraged to look up Hegel, Kant, and Spinoza on Wikipedia to learn what they taught so one can appreciate the ideas being absorbed by Dewey and the men of his time. Dewey's biographer reported the effect that these philosophy studies had on him: "Brought up in the tradition of liberal Congregationalist evangelicalism, he had at first no trouble accepting its teachings; later he found it increasingly difficult to reconcile certain of its doctrines with ideas he felt intellectually entitled to hold."

**On to the University of Michigan**

At the end of two years Dewey wrote his doctoral dissertation on "Kant's Psychology," got his Ph. D. and a teaching fellowship at Hopkins. As the protégé of the visiting professor from the University of Michigan, George Morris, who taught him at Hopkins, Dewey got appointed as an instructor of philosophy at Michigan. At that time professors of philosophy, usually clergymen, used philosophy primarily to support a theological position. But with Dewey at his side, Department Chairman Morris "introduced a new spirit of freedom in teaching philosophy." Morris emphasized the Germans, especially Hegel but the students were more radical than the professors. They claimed that Morris avoided British philosophers John Stuart Mill (atheist) and Hebert Spencer (agnostic) and the whole modern school of philosophy in order to not inflame the growing skepticism and agnosticism among the students. The students wanted instruction in the teachings of Mill and Spencer, especially their philosophy of religion. Spencer was the most famous European intellectual in the 1880s. Spencer developed an all-embracing conception of evolution as the progressive development of the physical world, biological organisms, the human mind, and human culture and societies. He was an enthusiastic exponent of evolution and wrote about it before Darwin did. The basis for Spencer's appeal to many of his generation was that he appeared to offer a ready-made system of belief which could substitute for conventional religious faith at a time when belief in the anchorless British religious sects was crumbling under the claims of modern science.

While at Michigan, Dewey became active in the local Congregationalist church, married a former student, and published his first book, *Psychology,* that was adapted as a text at Vermont, Williams, Brown, Smith, Wellesley, Minnesota, Kansas, and Michigan. A second book followed and won praise from faculty at Yale and the University of Chicago. Based on his growing reputation, he was appointed Professor of Mental and Moral Philosophy at the University of Minnesota in 1888. But in 1889, Professor Morris died and Dewey was offered the Chairmanship of the Philosophy Department at the University of Michigan, to which he returned. His first appointment was the son of a Congregationalist minister and a graduate of the Yale Divinity School. The next appointments were to Congregationalists who had studied at Harvard and in Germany.

### They Formed the Minds of the Elite

If the reader is finding this boring, perhaps a reminder may help. I am attempting to explain the religious and educational background of the faculties that shaped the thinking of the university students, the children of the well-to-do "movers and shakers" and the intellectuals in the late 19th Century and early 20th Century. This shaping of thought and the spreading of

Humanist philosophy explains why Humanism became dominant. The corollary is that belief in evolution became the norm because it is the base upon which Humanism rests. No Incarnation, no Christianity; no evolution, no Humanism. When Catholics accept belief in evolution it is not because they have thoroughly investigated the scientific claims. It is because they have accepted the testimony of Humanists regarding their base dogma. Catholics may have received that testimony from a sincere Catholic who received his belief from a previous sincere Catholic but the ultimate source of the testimony was a Humanist. The way that Humanists hired Humanists in the universities also helps explain the homogenous thought of university faculties on the subject of evolution.

Today, the "rule of science" enforced in university faculties and by professional organizations such as the American Association for the Advancement of Science is known as "methodological naturalism" or "methodological materialism." Stephen Meyer explained the rule as follows:

> Methodological naturalism asserts that to qualify as scientific, a theory must explain phenomena and events in nature—even events such as the origin of the universe and life or phenomena such as human consciousness—by reference to strictly material causes. According to this principle, scientists may not invoke the activity of a mind or, as one philosopher of science puts it, any "creative intelligence."

## Dewey Breaks from Religion

During his time at Michigan, Dewey began to shift away from Hegelianism because of the influence of psychology and evolutionary biology in his reading. He was very influenced by philosopher and psychologist Harvard Professor William James who was also the son of a Protestant theologian. James' writing about psychology influenced Dewey's opinions. Dewey began to see Christianity mostly in social terms and declared that "democracy, rather than the church, is the means by which the revelation of the truth is carried on." When Dewey left Michigan in 1894 to head the Philosophy Department at the University of Chicago, his connection with the Congregational church finally ended in a formal withdrawal of his membership.

At Chicago, as he had done at Michigan, he stacked the philosophy department with like-minded intellectuals. For example, he immediately appointed a faculty colleague from Michigan and a former student. The next appointments were U. of Chicago students who stayed on to get their doctorate in his philosophy department and then were made part of the philosophy staff. These colleagues and Dewey's students created a Humanist ripple effect over the years as they spread out and, in turn, became important faculty members in other colleges. Dewey had gifts of personal mannerisms that endeared him to students and colleagues alike. While at Chicago, one of Dewey's books was *Studies in Logical Theory.* William James wrote that it put into the world a view of the world, both theoretical and practical, which is so simple, massive, and positive that it deserves the title of a new system of philosophy. As noted at the beginning of this chapter, there is a Humanist world-view and a Catholic world-view. In 1897, Dewey gave a lecture to U. of Chicago students titled "Evolution and Ethics" in which he explained that "the ethical has its roots in the cosmic and is continuous with it."

### Dewey Begins to Influence Education Policy

In 1896, Dewey began giving lectures to Chicago's teachers. Dewey's impact on the teachers of that era was as great as his impact on philosophy. In 1899, he wrote *The School and Society* and it was gobbled up by leaders in education and teachers. It became a best-seller, went through seven printings in the next ten years and was translated into every major language.

Dewey's reputation in philosophy, psychology, and education resulted in guest lectures at other universities. He was also invited to speak at meetings of professional societies of educators. In 1888, he was elected president of the American Psychological Association. He was also active in the American Philosophical Association and became its president. In 1909, Dewey was elected VP of the American Association for the Advancement of Science. (To this day that organization is noted for vituperative abuse of creationists.) In 1915 Dewey was co-founder of the American Association of University Professors which, among other things, helped dissident theologians win the coup at Catholic University in 1967-69. In 1902, Dewey became director of Chicago University's School of Education and this eventually led to a dispute with the University's president, so Dewey resigned.

### Dewey at Columbia University

Dewey's greatest impact on America and on the world began when he was hired by Columbia. Dewey's appointment came with a seat in the Faculty of Philosophy, but the Faculty of Columbia's Teachers College arranged for him to be part of the Faculty of the Teachers College also. When Dewey joined Columbia it was already one of the nation's prestige universities. It also was among those with the largest faculty and student body. When Dewey went there in 1905, it had 5000 mostly local students. By 1930, it had nearly eight times as many and they were from all over the world. (If these numbers seem small by modern standards remember that in those days students or their parents paid their own expenses and the U.S. population was smaller.) With the addition of Dewey, Columbia's philosophy department was the leading university in Humanist philosophy.

And Dewey, through his association with the other "top names" in their field, imbibed more Humanist ideas. For example, a colleague in the Philosophy Department was Felix Adler, discussed earlier in this chapter as the founder of the Ethical Culture non-theistic religion. The Department Chair, Woodward, had planned to be an Episcopal Church clergyman but dropped out and went to study German philosophy. Another member of the Department, Montague, got his doctorate at Harvard and specialized in "speculative philosophy." During his first ten years at Columbia, Dewey published numerous articles in the *Journal of Philosophy,* and three books: *The Influence of Darwin on Philosophy and Other Essays, How We Think,* and *Ethics.*

**Dewey's Teachers Spread Humanism**

Columbia Teachers College was recognized as the home of liberal educational thought. Columbia Teachers College grew from 450 students in 1897 to 2500 by 1917. According to an article by Fr. John Hardon, written in 1952, there were 9,032 students in 1950. Teachers College was training twice as many teachers and educational administrators as any other institution in America. With a degree from the preeminent College, those students went out and obtained the most sought-after positions in influential schools and colleges. And they took Columbia's Humanist philosophy with them.

Dewey at Columbia often became involved in public education policy. For example, early in the 20th Century as public schools replaced church-affiliated schools, many leaders were concerned that most children attending public schools received no religious training. They were concerned that without religious training their moral lives and society would suffer. Proposals were made to release children for an amount of time during the regular school day for instruction in religion. The proposed plan was that they would be turned over to teachers from their respective churches or denominations for instruction. Advocates said that it could be put into operation without expense and it respected sectarian differences. Humanists opposed letting anything escape from their control. Humanists alleged that the public schools had been successful in assimilating different ethnic and cultural groups and that the "released time" would segregate them into denominational groups and cause them to notice religious differences of which they were not otherwise aware. Pope Leo XIII had identified removal of religion from school as a strategy of Naturalists and Freemasons.

Dewey threw his weight against the "released time" idea. As Dewey and Humanists in general had no belief in God they claimed that large numbers were finding the ideas and practices of traditional faiths inadequate. Speaking more for himself and his philosophy and psychology- immersed acquaintances who never received sanctifying grace, Dewey said people were searching for a religion that would more fully satisfy the moral, intellectual, and religious needs of the present. Dewey's conviction was that the faith of the future would center on the ideals of democracy and the findings of science. He urged that such a faith be taught in the public schools. He got his wish because that faith is taught namelessly in today's public schools. His biographer quoted him from an article Dewey wrote in 1908:

> Bearing in mind the losses and inconveniences of our time as best we may it is the part of men to labor persistently and patiently for the clarification and development of the positive creed of life implicit in democracy and in science, and to work for the transformation of all practical instrumentalities of education until they are in harmony with these ideas.

That sounds like what a more famous religious leader once said: "Go therefore and make disciples of all nations…" In the United States, Dewey's disciples have out-evangelized His disciples. That was the way Dewey saw that public education could be used to evangelize for the religion of the future, with that religion being formally announced in the publication of The *Humanist Manifesto* in 1933 that Dewey signed. A new non-theistic religion was a goal that Dewey preached throughout his long career at Columbia that ended in 1939 when he was 80, and which he continued to advocate in his retirement years. For example, in 1940 the question of released time for the New York City schools came up again. The Board of Education held a meeting that parents, teachers, civic, and religious leaders attended. Dewey, representing the Committee for Cultural Freedom, an atheist front group of which he was honorary chairman, spoke against released time. He cloaked his ideological opposition to theistic religion by wrapping it in his revisionist version of the intentions of the Founding Fathers:

> I do not think that the men who made the Constitution forbade the establishment of a State church because they were opposed to religion. They knew that the introduction of religious differences into American life would undermine the democratic foundations of the country.

Certainly Dewey knew that at the signing of the Constitution some of the 13 states had established religions and banned certain religions so the 1st Amendment prohibited Congress from establishing a national church and had nothing to do with their alleged belief that religious differences would undermine the Republic. Dewey's biographer explained Dewey's real motive.

> It is better for the schools to continue following along the lines they had been following, Dewey believed, than that

they should, under the name of spiritual culture, form habits of mind that are at war with the habits of mind congruous with democracy and with science.

### Worldwide Evangelist

Dewey's reputation was such that he was in demand around the world as his books were translated into all of the major languages including Arabic, Chinese, Japanese, Turkish, Persian, and Portuguese. On a sabbatical leave from Columbia in 1918-1919, his wife planned a vacation trip to Japan. When they heard he was coming, a Japanese man who had met Dewey when getting his Ph. D. at Michigan arranged for Dewey to give a series of lectures at the Imperial University in Tokyo, and a thousand people turned out for the first lecture. His theme was the same, democracy and science had replaced the old values. When the Chinese learned Dewey was in Japan, he was invited to lecture at the National University in Peking (now Beijing) for the academic year 1919-1920. He gave lectures that were simultaneously translated into Chinese, recorded, and then transcribed for the press and scholarly journals. Many were published in book form in Chinese. He was urged to stay on for the academic year 1920-1921 as a visiting professor of philosophy and got approval from Columbia to do so. Dewey taught at the National University and the National Teachers College. During his time in China, Dewey spent some time with the famous, atheist British philosopher Bertrand Russell, and their debates attracted Chinese intellectuals, including the young Mao Tse-tung, who killed millions when the Communist Party he led came to power after World War II. Dewey toured China giving lectures all over. Wherever the Deweys went they were given a big welcome party, attended by government, civic, professional, and educational officials.

Back at Columbia in 1921, Dewey continued writing, teaching, and evangelizing. In the summer of 1924, he was invited by the Turkish Government to survey the country's educational system and recommend ways to improve it. In the summer of 1928, he was part of a delegation invited by the Communist Soviet Commissar for Education to visit schools in Leningrad and Moscow. What attracted Dewey most about the schools was that they were made to serve the needs and interests of a Communist society by teaching evolution and atheism. John Dewey retired in 1935 but was a professor emeritus until 1939. He continued writing and speaking and influencing for the doctrines of Humanism for another 10 years.

### Prophet of American Naturalism

That is what Fr. John Hardon called Dewey. To repeat, it was not to bore the reader that so much has been written here about John Dewey. It was to make sure that the reader understands that most of our American universities and their leading intellectuals from the beginning have been of heretical religious background, or no religious background, and deeply immersed in a philosophy and a cult of science that has no room for God as Catholics know Him.

### Catholics Once Resisted but Their Intellectuals Capitulated

It was not as if Dewey and the "progressives" found no opposition, at least at first. Traditional Christians strongly supported science and the scientific method in the curriculum but those subjects should not take over roles properly belonging to the inductive and deductive methods of acquiring truth. The Humanists insisted that the "wisdom of past generations" should be viewed only as tentative in their application to the present. They promoted the principle that free, open critical discussion is at the heart of political democracy and must also be the principle pervading American education. The latter is a tried and true technique of Humanists, namely, to preach free, open critical discussion and tolerance until they get in the position to suppress all viewpoints but their own. They have achieved that position in American public and education policy today.

### Catholics Were Once Awake

What Dewey and his like-minded Humanists were doing to education was recognized for what it was and, believe it not, Catholics were actually awake in the 1940s and providing opposition before many of their clergy succumbed to evolution, such as when Fr. Bruce Vawter was a seminarian in 1946. While continuing their criticism of his Humanist takeover of public education, Dewey's opponents attacked his overall philosophy of experimental, humanist naturalism that was adversely affecting the lives of students. They rightly saw how deeply this philosophy was entrenched in the universities and even the elementary and secondary schools. Fr. Geoffrey O'Connell, speaking at the National Catholic Alumni Federation meeting held in New York in 1939, pointed out that, for more than three decades, John Dewey and his followers at Columbia's Teachers College had made that institution the center of their operations in their "attempted destruction of Christian aims and ideals in American education." *The Tablet*, a newspaper published by the Diocese of Brooklyn, NY, denounced Dewey's philosophy three times in 1939-40. Fr. O'Connell and other Catholics, it appears, were seeing firsthand and resisting the methods of the Naturalists and Freemasons described by Pope Leo XIII.

### Human Rights Abuse Predicted

It was also recognized in 1939 that Dewey's philosophy would lead to abuses of human rights because human rights and human worth can be respected only when it is recognized that they are supernaturally derived. Dewey taught that the dignity of man and human rights have evolved naturally in the course of man's social, moral, and cultural advances. It never occurred to Dewey, one supposes, that the culture of slavery and death that was the norm in the Greco-Roman Empire before Christianity is what actually evolved naturally in the course of man's social, moral, and cultural advances. The culture into which Dewey was born and prospered was a decaying but still living culture built by the Catholic Church. Even some modern atheist writers have admitted that they have been "freeloaders" on the inherited culture. Certainly with "democracy and science" as the basis for his non-theistic religion, Dewey would have to accept that whatever future course the dignity of man and human rights might take would be legitimate advances. We can judge that he would have approved of *Roe V. Wade* based on his being a signer of *The Humanist Manifesto I.* Certainly he would be a defender of Planned Parenthood's selling of carefully butchered babies if he were alive today.

Speaking at the Columbus Forum in December 1939, William Parsons pointed out that in denying the supernatural basis of human rights, Dewey's philosophy was identical to that which was the basis of the three forms of totalitarian dictatorship then strangling Europe: Communism in Russia, Fascism in Italy, and National Socialism (Nazism) in Germany. Such ideologies held, with Dewey and Humanists, that man is nothing more than a highly developed animal, with no rights other than those conferred by government or society. That's pretty much the philosophy of those running the United States' "soft tyranny," which bends the Constitution at will.

**Schools Used As Tools**

The opponents feared then what has become a reality today, namely, that the classroom would be used to undermine America's faith in its founding principles of inalienable rights articulated in the Declaration of Independence and the Bill of

Rights Amendments to the Constitution. Thomas Woodlock, addressing the National Council of Catholic Women Convention in 1939, declared that it is in the educational system in this country that there lies the danger of totalitarianism in the clothing of democracy.

Mortimer Jerome Adler was a philosopher, educator, and popular author who got his doctorate in psychology at Columbia. As a philosopher, he once taught at the U. of Chicago but in the Law School because the staff Dewey left behind in the philosophy department when he went on to Columbia, all Humanists, opposed Adler's appointment to that department. Although he was a nonobservant Jew, he was attracted to the work of St. Thomas Aquinas. He may be best known for the Great Books of the Western World program that he co-founded. Adler addressed the First Conference on Science, Philosophy, and Religion in New York in 1940. He was very prescient in speaking about the professorial profession, which Dewey helped shape when he organized the American Association of University Professors. He attacked the professorial profession. He said that democracy has much more to fear from the mentality of its teachers than from the nihilism of Hitler. He said it is the same nihilism in both cases, but Hitler's is more honest, less blurred by subtitles and queasy qualifications, and hence less dangerous. Ouch!

## Control the Textbooks

Progressives managed to get control of the textbooks. Texts recommended more government planning and control in areas such as business and industry, health care, housing, and rehabilitation frequently citing the Soviet Union and Scandinavia as models of nations where social planning was accepted as a proper function of government. A faculty member at Teachers College produced a textbook for school children that was very effective in making socialists of them. It was adopted for use by

4000 school systems. Today all of the public school textbooks promote ideas in harmony with the Humanist religion.

Whatever concern Catholic clerical leadership in America once had in protecting Catholic children in public schools from Humanist indoctrination seems to have disappeared. Many of the clerical and lay intellectuals, such as philosophy and theology faculty lacking the diligence to investigate the science, created a new man-made religion called 'theistic evolution,' which is neither Biblical nor scientific. They claim that by attributing to God the trillions of supernatural interventions necessary to make things evolve they are actually giving God more credit than *Genesis* gives Him. Humanists don't mind those intellectuals because they stay behind the doors of their churches and academies and leave the public schools to Humanists to make it plain that no God is necessary or desirable.

**What Parents Are Up Against**

It is up to parents to arm their school children. It is not easy for children who have been educated about creation to get along in public schools because disagreement with the teacher can result in retribution against them. A good example of how Humanist "temples," otherwise known as universities, evangelize and intimidate is an article called "Darwin Defended" that was reprinted online at Slate.com in March 2015. It was written by a religiously-zealous promoter of evolution, a biology teacher at the University of Kentucky. He has been teaching evolution to 1800 students a year for 20 years. The students he teaches are non-biology majors for whom his class is a freshman requirement. His theme in his article was that evolution is 100% scientific but there are some students he'll never reach because they are backward Evangelicals and Catholics who have dared to question his doctrinaire approach.

> The story of our evolutionary history captivates many of my students, while infuriating some. During one lecture, a student stood up in the back row and shouted the length of the auditorium that Darwin denounced evolution on his deathbed—a myth intentionally spread by creationists. The student then made it known that everything I was teaching was a lie and stomped out of the auditorium, slamming the door behind him. A few years later during the same lecture, another student also shouted out from the back row that I was lying. She said that no transitional fossil forms had ever been found—despite my having shared images of many transitional forms during the semester.

As this writer has explained in this book, evolutionist teachers are still marketing "images of many transitional forms" when all of the professional paleontologists have accepted that there aren't any transitional forms. In his 1980 book, *The Panda's Thumb*, Gould called the absence of transitional forms "the trade secret of paleontology" which he had tried to explain by punctuated equilibrium before he eventually recanted. To the Catholic students, the U. of Kentucky teacher tells the following lie:

> Even Pope John Paul II acknowledged the existence of evolution in an article he published in the *Quarterly Review of Biology*, in which he argued that the body evolved, but the soul was created.

**Liar or Ignorant**

Pope John Paul II never published an article in the *Quarterly Review of Biology*. Someone wrote an article in that journal about the Pope's address to the Pontifical Academy of Sciences and if that person wrote that the Pope "argued that the body evolved," he is a liar also. The Kentucky teacher whined about the few students who "aren't buying" in the same way that

Stephen Jay Gould whined in his introduction to *evolution: Triumph of An Idea,* which was discussed earlier.

> We live in a nation where public acceptance of evolution is the second lowest of 34 developed countries, just ahead of Turkey. Roughly half of Americans reject some aspect of evolution, believe the Earth is less than 10,000 years old, and that humans coexisted with dinosaurs. Where I live, many believe evolution to be synonymous with atheism, and there are those who strongly feel I am teaching heresy to thousands of students.

The teacher is just plain exasperated by those intelligent Kentucky Evangelicals.

> After a semester filled with evidence of evolution, one might expect that every last student would understand it and accept it as fact. Sadly, this is not the case. There are those who remain convinced that evolution is a threat to their religious beliefs. Knowing this, I feel an obligation to give my "social resistance to evolution" lecture as the final topic. This lecture lays down the history of the anti-science and anti-evolution movements, the arguments made by those opposing evolution, and why these arguments are wrong.

### Intellectual Child Abuse

This teacher wouldn't last 5 minutes in a debate with an adult who knew something about this subject. But he doesn't have to debate. He has a pulpit and he uses it while Catholic pulpits are silent on this subject. What's going on at the U. of Kentucky and in most other universities is intellectual child abuse.

The article contains a vignette about the "Scopes Monkey Trial" which, to the teacher, was a great triumph of his heroes standing up for "science." He probably wows the freshman with the phony glory of John Scopes and Clarence Darrow. He doesn't mention what has been explained earlier in this book, namely, that the "scientific evidence" provided by his evolutionary heroes included the fraudulent "missing link" Piltdown Man and "scientific" testimony about 180 human body parts that 19th Century science had "shown" to be left over as humans evolved from lower ancestors. He quoted Stephen Jay Gould who explained how theories are facts. (That's Gould's semantic "shell game" of substituting "fact" for "data" as explained in the next chapter.)

For those still asking "how can evolution be false when everybody who is not a backward creationist believes it?" the following part of the teacher's story may help to explain:

> I was originally reluctant to take my job at the university when offered it 20 years ago.... I wasn't particularly keen on lecturing to an auditorium of students whose interest in biology was questionable given that the class was a freshman requirement. Then I heard an interview with the renowned evolutionary biologist E.O. Wilson in which he addressed why, as a senior professor—and one of the most famous biologists in the world—he continued to teach nonmajors biology at Harvard. Wilson explained that nonmajors' biology is the most important science class that one could teach. He felt many of the future leaders of this nation would take the class and that this was the last chance to convey to them an appreciation for biology and science.

In other words, it provided the last chance to inculcate students who would never learn enough biology and science (or Christianity) to ever question what they learned in that class and thus ensure that they would become life-long evolutionists open to the Humanist philosophy. Did any of Wilson's students, "future leaders of this nation", leave Harvard <u>not</u> believing evolution was a fact? At the U. of Kentucky, 1800 such students are evangelized each year. Multiply 1800 a year by the number of universities teaching the same or similar and it helps explain the statistic given in Chapter Two that 30% of those who lose their Faith lose it between 18 and 23.

### Other Popular Evolution Propaganda

In the late 19th century Humanists cranked out books such as *History of the Conflict Between Religion and Science* (1874) and *A History of the Warfare of Science with Theology in Christendom* (1896). As Jonathan Wells correctly noted in *Zombie Science*, "there never was a war between religion and empirical science, but there is a war between religion and materialistic science and the battleground in evolution." In the early 20th Century, it was not just the philosophy professionals like Dewey who were exerting a push toward a Humanist worldview. British science fiction writer H.G. Wells is best remembered for books such as *The Time Machine* (1895), *The Island of Doctor Moreau* (1896), *The Invisible Man* (1897), and *The War of the Worlds* (1898). Wells had a B.S. in zoology and his thinking was definitely Darwinian. Twenty years after his last science fiction best-seller, Wells took up writing fictional history. *The Outline of History* was originally published as a series in the British Humanist periodical, *The Fortnightly*. It was published in two volumes in 1920. As history, the professional historians weren't impressed, but it was popular with the general public. It sold 2 million copies and made Wells rich. We have Wells to thank for G.K. Chesterton's brilliant work, *The Everlasting Man,* published in 1925. In refuting Wells, from the very first pages, Chesterton attacked evolution as a science and a philosophy that is both improvable and implausible. He did so with logic and wit throughout the book because evolution as a fact was the basis of the Wells book. It is puzzling to this writer that so many Catholic theistic evolutionists, who so greatly admire Chesterton and *The Everlasting Man,* don't get the connection that evolution is the basis of a philosophy which is diametrically opposed to Catholicism. Chesterton certainly did. And so did C.S. Lewis who is also highly admired by orthodox Catholic intellectuals.

## Pop Philosophy

In this country the noted historian, evolutionist, Humanist, and anti-Catholic Will Durant contributed to promotion of Humanism with his *The Story of Philosophy* published in 1926. Durant started with the pagan Greeks and then skipped forward to the 16th Century. Beginning there, he covered all of the Humanists philosophers up to, and including, John Dewey, his contemporary. According to Durant, the Christian philosophers, i.e., the ones who founded and taught at the great universities that brought Europe out of the Dark Ages caused by the invasion and destruction of the Greco-Roman culture by barbarian hordes from northern Europe, weren't worth considering. He explained in the preface to the second edition that "the total omission of scholastic philosophy" was because he had "suffered much from it in college and seminary and resented it thereafter as rather a disguised theology rather than an honest philosophy." The first edition of Durant's book was translated into German, French, Swedish, Danish, Chinese, Japanese, and Hungarian. The American edition sold 650,000 copies. The book was intended to make philosophy available to the non-college public. According to Durant, sales of philosophy classics (by that he means of agnostic or atheist authors) increased 200 per cent. "Many publishers have issued new editions of Plato, Spinoza, Voltaire, Schopenhauer and Nietzsche." He said that an official of the New York Public Library told him that ever since the publication of *The Story of Philosophy* the library had a wide and increasing demand for the philosophical classics. Thus it was that Humanist philosophy was brought to John Q. Public.

## Pop Psychology

Another example of how belief in evolution was indirectly communicated to the American public is *The Road Less Traveled.* Written in 1978 it eventually spent 13 years on the *New York Times* bestseller list to create a paperback record, sold 10 million copies worldwide and was translated into more than 20 languages. It was written by a psychiatrist with no particular interest, it seems, in promoting evolution. While he obviously believed it, he called it a miracle because he admitted it conflicted with the Second Law of Thermodynamics which is that things tend to disorder over time, not to become more ordered. The author believed in it because that's what he was taught to believe. He devoted 5 pages to the flow from viruses to bacteria to fish to birds to animals to man according to the typical evolution story.

> The process of evolution can be diagrammed as a pyramid, with man, the most complex but less numerous organism, at the apex, and viruses, the most numerous but least complex, at the base.

If ten million copies were sold, even though the book was not about evolution, a lot of people who might not otherwise have thought much about evolution were told it was a fact.

## Pop Cosmology

Catholic theistic evolutionists continue to doubt fiat creation and ask why the majority of Americans believe in evolution if it is not true. Many reasons are given in this book but this is a good place to cite the role Carl Sagan played in helping to convert today's adults. Like Stephen Jay Gould, the late Carl Sagan was the New York city-born Humanist son of nonobservant Jews from Eastern Europe. It would take many pages to describe all of Sagan's contributions to Humanist propaganda but anyone interested can look him up on the internet.

Sagan is most famous for a 13-part PBS series that he co-wrote and presented called "Cosmos: A Personal Voyage." The 13 episodes were broadcast by PBS in the autumn of 1980 at which time Dr. Consolmagno, the Pope's astronomer, was a lecturer at the Harvard University Observatory. "Cosmos" was the most widely watched series in the history of American public television until 1990. As of 2009, it was still the most widely watched PBS series in the world. It won two Emmys and has been broadcast in more than 60 countries. It is estimated that over 500 million people have seen it. A book, found in my local public library, was also published to accompany the series. Subsequently Humanist, anti-Christian Ted Turner bought "Cosmos" from PBS, with Sagan he expanded it, turned it into a commercial movie and DVDs in boxed sets are sold around the world.

Carl Sagan, like most of the other famous Humanist evolutionist celebrities such as Richard Dawkins and Stephen Hawking, was an avid believer in extraterrestrial life. His belief, like theirs, resembled the PBS "Origin of Life" description of life "arising" from space debris that was described in chapter four. One can get the flavor of the ideology of Sagan's PBS series from Episode One, "The Shores of the Cosmic Ocean." Sagan opened the program with a description of the cosmos and a "Spaceship of the Imagination" (shaped like a dandelion seed.) The ship travels through the universe's (supposed) hundred billion galaxies, the Local Group, the Andromeda Galaxy, the Milky Way, the Orion Nebula, our Solar System, and finally the planet Earth. Evolutionists dazzle viewers with such audio-visuals because, as explained in chapter five, the origin of first life is their Achilles Heel. The only way to avoid science is to substitute science fiction.

By the second episode he “explained” evolution through “natural selection” (and the alleged pitfalls of intelligent design). He resurrected the “Primordial Soup Hypothesis” by stating that the Miller-Urey experiment of 1952 (described for what it was in chapter four) demonstrated the “creation of the molecules of life.” From there he moved on to speculation about alien life in Jupiter’s clouds. Anyone who wants to learn what else the gullible people of that generation learned from “Cosmos: A Personal Journey” can look it up on the internet.

### More Humanist Propaganda for This Generation

In 2010 the Discovery Channel televised the 3-episode “science documentary” *Into the Universe with Stephen Hawking*. When it came to the origin of life on earth Hawking offered two atheistic evolutionary hypotheses in answer to what to him is a problem. These were the primordial soup theory and panspermia (life began elsewhere and was seeded to Earth by asteroids). The bogus science was dressed up with computer-generated imagery of the universe and an original soundtrack combining symphonic orchestral recordings with electronic and sampled elements.

In 2014 millions and millions of TV viewers saw on 21 Fox Network channels (and others internationally) “Cosmos: A Spacetime Odyssey.” This is a follow up to the 1980 TV series. It was co-written and co-directed by Sagan’s widow who also co-wrote the first cosmic fantasy. It was bankrolled by the man who created and owns the crude animation program on Fox, “The Family Guy.” The new “Cosmos” loosely follows the same 13-episode format and storytelling approach that the original “Cosmos” used, including elements such as the "Ship of the Imagination" and the “Cosmic Calendar.” To sell the storytelling it utilizes extensive computer-generated graphics and animation footage to augment the narration.

Watching video of evolution propaganda like "Cosmos" that relies on computer-generated graphics and animation as a substitute for reality always reminds me of the movie, "The Wizard of Oz." The "Great and Powerful Oz" has everybody fooled and intimidated until Toto pulls back the curtain and reveals that Oz is just an old carnival performer. The Discovery Institute pulled back the curtain on "Cosmos." So scientifically fictional was "Cosmos" that later in 2014 the Discovery Institute published a book called *The Unofficial Guide to Cosmos: Fact and Fiction in Neil DeGrasse Tyson's Landmark Science Series.* The book demonstrated that Cosmos is an agenda-driven vehicle for scientific materialism, casting religion as arch foe of the search for truth about nature and pressing its message that human beings occupy no special place in the universe. The book is an episode by episode review of where "Cosmos" veers from objective science to science-flavored, fact-challenged preaching. But "Cosmos" headed straight into schools as a science teacher's instructional aid and the kids will never hear the truth. Neither will the millions of adults who watched it on TV, unless readers tell them.

**Historical and Religious Fiction**

Along with science fiction "Cosmos" featured historical and religious fiction. For example, in the opening episode, riding his "Ship of the Imagination," the host and narrator, Neil DeGrasse Tyson, took the viewer to show where Earth sits in the scope of the known universe. He then explained how humanity has not always seen the universe in this manner, and described the hardships and persecution of Renaissance Italian Giordano Bruno in challenging the prevailing geocentric model of the universe (supposedly) held by the Catholic Church. Bruno was burned at the stake in 1600. This story is a variation of the Humanist tactic described in chapter 13 and Appendix I regarding Galileo. Trot out somebody to "prove" the Catholic Church is the enemy of science. It required more Humanist lies than usual to make Giordano Bruno a martyr for science. While it is true that Bruno supported the heliocentric model of the universe according to Copernicus, Bruno was not scientifically important in his own right. The Humanists writers implied that Bruno was burned for support of the heliocentric theory. In 1593 when Bruno's trial began, the heliocentric theory was not a scientific issue. There was a very strong scientific consensus behind the geocentric model. The first published defense of the Copernican Model was Johannes Kepler's 1595 work, *Mysterium Cosmographicum* (*The Cosmographic Mystery*). Galileo, although he seemed to believe in the heliocentric model, taught the geocentric model at the University of Padua from 1592 to 1604 for fear of being ridiculed. Besides that, the Catholic Church has never "held" the geocentric model as a doctrine although it was the scientific consensus until at least 1687. What the anti-Catholic, Darwinist writers "forgot" to mention was that the Dominican priest Bruno was actually convicted as a denier of the Trinity, the divinity of Jesus, Mary's virginity, transubstantiation, and he was a pantheist. It had nothing to do with science.

Unrepentant prominent Catholic heretics like Bruno were sometimes burned in Catholic-ruled countries. In merry old England people were hanged, drawn and quartered just for being Catholic. (If you don't know what that involved, check it out on Wikipedia.) In 1741, in British-ruled New York City, 13 Negros were burned at the stake and 17 others were hanged. We are more civilized now thanks to our Humanist-controlled culture. We only permit babies to be killed, cut up and sold for "research" as we go about our normal highly-cultured lives.

### Intellectuals and the Humanist Manifesto of 1933

In his book, *The Genesis of a Humanist Manifesto*, Edwin H. Wilson, one of the founders and preeminent leaders of the

Humanist movement, explained how evolutionary cosmology and biological evolution undergird Humanism. Wilson wrote that "Humanism came of age in 1933 with the publication of Humanist Manifesto I." Wilson stated that its affirmations of faith regarding cosmology, biological and cultural evolution, human nature, epistemology, ethics, religion, self-fulfillment and the quest for freedom and social justice described precisely "the leading ideas and aspirations of its era."

**Not an Idle Boast**

That was no idle boast. As one ought to have learned after reading the life and times of John Dewey in this chapter, Edwin Wilson correctly observed that *The Humanist Manifesto* reflected the reality that by 1933 "what was conceived by the convergence of freethought and religious liberalism at the end of the Nineteenth Century" had come to reign in the universities, if not yet in the local school houses. As has been explained, Humanist philosophy anchored on origin speculations is as old as the famous Greeks of the pre-Christian era, but it lacked a basis to plausibly refute the Bible's Divine Revelation. Humanist philosophers in Europe labored mightily to supersede the scholastic philosophy of Church Doctors, who developed it in harmony with the Magisterium. Two books of a scientific gloss, written by Lyell and Darwin, which have been discussed in earlier chapters, appeared in the Nineteenth Century and pumped new life into Humanism by advancing theories to explain observed data.

Anyone reviewing the list of that *Manifesto's* signers might wonder "so who were they?" and not notice the name of John Dewey, the most influential philosopher in American history. At this point, my objective is to state just the first 3 of the 14 affirmations of faith of Humanists in support of my assertion that Condition #2 of those stipulated by Pius XII for any research and discussion of evolution could never be acceptable to the Humanist scientific consensus. The first three are:

> Religious humanists regard the universe as self-existing and not created.
> Humanism believes that man is a part of nature and that he has emerged as the result of a continuous process.
> Holding an organic view of life, humanists find that the traditional dualism of mind and body must be rejected.

In other words, the universe including all of the matter and energy always existed, life "emerged" on its own, humans then evolved, and the mind is material just as the body is. No room here for spiritual souls.

**Adult Self-Education Time**

Readers are urged to inform themselves about Humanists beliefs and the Humanists political plan for the world. On the internet, search for www.americanhumanist.org which is the home page of the American Humanist Association. See the Humanists current agenda for public policy. Notice how involved Humanists are in advocating public policy while Catholics are primarily just reacting to the latest outrage. In the very upper right of the home page there is a "search" box. Type in *Humanist Manifesto I,* print it out and study it. Type in *Humanist Manifesto II.* Print and read it. Keep in mind when reading *Manifesto II* that it was written in 1973. In 1973, Humanists articulated the way things ought to be. Try to identify one of the goals found there that has not become either enforced by law or enforced culturally by "political correctness." Christians who have been wondering about the ever-creeping assault by Government and the culture on their basic values might recognize that the program laid out by Humanists in 1973 is now in place. Humanists have no intention of peaceful co-existence with Catholics or others who take the Bible seriously. They have no tolerance for any worldview but theirs. You have wasted your time reading this book if you will not go online and read those Humanist documents otherwise you may think that the decline of morality in American culture and government's creeping tyranny toward Christians "just happened" naturally.

At the beginning of this chapter I said the purpose of this chapter is to answer this question: "If the theory of evolution is just so much scientific bunk as you have said, how can you explain that it has become accepted as the scientific consensus and sixty-five percent of American adults believe it?" If my explanation was convincing no reader should continue to ask how evolution came to be so widely accepted in America despite its scientific foibles.

### Humanism Includes a Political System

Christians in this country are already subject to a "soft tyranny" at many levels such as public education. If one thinks it is not going to get worse, he hasn't understood the "religious" zeal of Humanists to impose their worldview despite a temporary setback in the elections of 2016. Trump was not elected by the elite. The graduates of the elite universities who run this country can relate more to the Creed of the *Humanist Manifestoes* than to your Creed. And they are not going away. They are the "deep State." News articles such as "Rogue Federal bureaucrats threaten Trump's agenda" discuss the reality of the permanent entrenchment that is the result of decades of Humanist college graduates gravitating to Federal employment. During the campaign 95% of the political donations from employees at 14 federal agencies went to Hillary Clinton, a Democrat-Republican donation ratio more imbalanced than the more notoriously partisan gap among university faculty.
http://thehill.com/homenews/campaign/302817-government-workers-shun-trump-give-big-money-to-clinton-campaign

A *Washington Post* article "Resistance from within: Federal workers push back against Trump" explained:

> Less than two weeks into Trump's administration, federal workers are in regular consultation with recently departed Obama-era political appointees about what they can do to push back against the new president's initiatives. Some federal employees have set up social media accounts to anonymously leak word of changes that Trump appointees are trying to make.

A week later the *Washington Post* published another article with the same theme: "Staying true to yourself in the age of Trump: A how-to guide for federal employees."

Catholics must learn to push back against Humanists. Just prior to the reprieve granted by Trump's election some Church leaders had been "handwringing" and telling the laity to prepare spiritually for a coming persecution. In this window of opportunity Church leaders ought to speak against the Humanist resistance to whatever pro-Christian reforms Trump may try to initiate by calling it what it is. Humanism is an oppressive religion and must be opposed as a religion by religious truth. Sadly, many Catholic bishops, such as the Bishop of San Diego, openly joined the Humanist Left against the President. The laity must also push back against rogue bishops and clergy.

**Catholic Malaise**

Michael Voris of St. Michael's Media commented on the vague discomfort that he sensed prevailed among Catholics in July 2015. Most of what he wrote still applies:

> If you haven't noticed, there is a general feeling of malaise going around the Catholic Internet these days, brought on and exacerbated by the U.S. Supreme Court's tragic decision making sodomy "marriage" a

> constitutional right. However, there is something of a counterbalance to the malaise also going around. It's the phrase "We know who wins in the end"... While that is true as it stands, it is has the potential to inspire many to non-action. It gives cover to those who don't really want to do anything because they are paralyzed by fear, or cowardice, or have just run out of gas. Allow [me]...to submit to you that this is a very dangerous attitude. It is also presumptuous. First, anything that excuses fear, coddles exhaustion, or vents righteous anger away from its justified target is bad. Second, it is presumptuous in that the person saying it can potentially assume that *he* is included in the "we" of "We win in the end." ...No Catholic can think the battle is just some passing thing, not really affecting them, and they can just wait for the reinforcements to come over the hill while they essentially "sit this one out." ... This is a fight to the death, and this is the point too many...are failing to grasp at this critical moment...It is totally insufficient to sit back and adopt the attitude: "We know how this ends; we win." It's almost a safe bet that many who take comfort in that saying will surely *not* be among the "we" they speak of.

Christians are spending hours reading about, blogging about, emailing or posting on Facebook stories about the outrages of the Humanist Left. What is needed is *organized action* to become educated on these matters and to start pushing back effectively.
That begins with pushing back against the public and private education system that through bogus "science" is alienating children from Christianity and turning them into Humanists.

# Chapter 15-Education Is Within Reach of All

There is no need for this writer to add to what Pius XII and Cardinal Ratzinger diagnosed as the sickness. The most important part of the cure was also prescribed by them, namely, for Church educators to rediscover and start teaching the authentic doctrines of creation and to stop polluting theology and philosophy with evolution. This is a massive long-term project that Catholic priests in communion with the Magisterium must accept and arm themselves to do. Every adult is capable of looking up and comprehending the difference between science and philosophical speculation. For example, when "evolution deniers" (as those who are of this writer's opinion are called) say "Evolution is not even a scientific theory" what do they mean? If one looks up the standard definition of a scientific theory he will find something very like this one found on Wikipedia:

> A scientific theory is a well-substantiated explanation of some aspect of the natural world that is acquired through the scientific method and repeatedly tested and confirmed through observation and experimentation. As with most (if not all) forms of scientific knowledge, scientific theories are inductive in nature and aim for predictive power and explanatory force.

The theory of evolution provides an explanation for events of the pre-historic past (billions of years ago it says) based on inferences made from presently observable data. Nothing about those events can be repeated or tested through observation and experimentation. Evolution cannot account for the origin of life, the sudden appearance in the fossil record of diverse fully-formed animals with no ancestors, or the information necessary for anything to become a more complex animal. It has no predictive power. All "acceptable" inferences must exclude anything but natural causes. Inferences are not proofs. Readers should now understand the speculative nature of evolutionary cosmology. Readers should now understand why "upward reasoning" from biological microevolution to macroevolution "invariably fails" as even the "Darwin of the 20th Century," Ernst Mayr admitted. Evolution as an explanation for anything is "dead in the water." Books and websites anyone can utilize have been cited. With a little effort one can learn how the observed data make more sense according to the creation model than according to the evolution model. With a little study one learns that evolutionist assumptions are used and treated as axioms. We know those assumptions are false, yet they use them anyway and hide them from those they teach.

## It's Not Obscure

One doesn't have to be a science professional to understand these concepts. Recall this writer's wife's testimony: homemaker mother of six, when challenged by the Headmaster at our son's school, picked up *The Genesis Flood*, read it, and became quite capable of defending creation. So much more creation-supporting science research has been done since 1961 when that book was published. When she wrote her own book in 2009, the title reflected her abiding Faith in God's Word. The title was *A Bird in the Hand*. In the book, she explained the title came from the saying "A bird in the hand is worth two in the bush." Catholics have in their hand something real and tangible, God's testimony regarding origins. The speculations of the Humanists regarding origins are the birds in the bushes, nothing but speculative conjectures.

## Simple and Factual Communication

The Barna Group is America's premier research organization devoted to studying cultural and religious trends. In 2011 Barna released the results of a 5-year study project under the title "Six

Reasons Young Christians Leave Church," which found that a main reason young people are leaving is that the church is not keeping up with or teaching people how to interact with science. Philosopher J. P. Moreland, a Fellow at the Discovery Institute's Center for Science and Culture, agreed but said

> The sad thing is that Christian scholars are, in fact, doing just this. The quality of Christian literature is getting better and better when it comes to showing that the Bible gets it right. Both theistic and naturalistic evolution theories are rationally inferior to Intelligent Design theory theologically, philosophically and scientifically. But people don't know this.

For their part, the laity and clergy can understand and communicate that the disagreement between practitioners of evolutionary science and creation-supporting science is not about the observable data. The disagreement is about the scientific interpretation of data because for them, who will admit of no God, interpretation and inference is all there can ever be regarding origins. That is, we are not arguing over the boiling point of water. Instead, we are arguing over interpretations of evidence that are couched in evolutionary terms, collected under evolutionary experiments, and explained to the masses by ardent evolution supporters. Evolution, then, is not a Humanist conspiracy so much as it is a prop in a mass movement away from God. In another sense, it is a smokescreen designed to mask a raging spiritual battle for human souls.

Learn the difference between "operational" and "historical" science. Operational science formally deals with what we can see in the laboratory *today*. It deals with repeating experiments, testing results, and refining hypotheses. It is the type of science that led to the development of our modern technology. And it has nothing to do with evolution or deep time.

Historical science, however, attempts to draw conclusions about one-off things that happened in the remote past. But history is not testable, neither is it repeatable, so studying the past is not operational science. This does not mean we can't know anything about the past but we need to be more careful when drawing historical conclusions.

Once someone comes to the conclusion that the majority of the world's opinion leaders are wrong about something, the next obvious question is, "*Why* are they wrong?"

For the reasons given in the last chapter, the answer is plain to see. The majority of moderns who work in or teach in the historical sciences have been taught that everything in the universe can be explained by natural causes. In practice, it becomes a demand that all things *must* be explained by natural causes, which by necessity excludes many theists from "the club" that oversees scientific employment, advancement, and publishing.

Of course, there are multiple gigantic problems that arise once someone makes that assumption, but belief in evolution is not inimical to the collection of most scientific facts. For that reason even those who work in or teach in the operational sciences may also accept naturalism because that is what they have been taught and, since evolutionary cosmology and biology have no practical connection to their work, they would have no reason to question what they were taught.

Included in the world's opinion leaders who are not scientists but who accept naturalism because that is what they were taught would be those non-biology majors of Harvard's E.O. Wilson and the U. of Kentucky professor discussed in the last chapter.

## Gould's Shell Game

Stephen Jay Gould who, as explained Chapter 12, made a faith choice when he was a child and stuck with that choice his whole life, is a classic example of a historical scientist whose writings have converted many to naturalism or reinforced what they had learned in school. In an article published in a 1982 book called *Speak Out Against the New Right* Gould provided what was an exercise in atheistic evolution apologetics that other atheists have repeated. In it he explained how "data" become "facts" when the starting point is that evolution certainly happened and just needs to be explained.

> Well, evolution is a theory. It is also a fact. And facts and theories are different things, not rungs in a hierarchy of increasing certainty. Facts are the world's data. Theories are structures of ideas that explain and interpret facts. Facts do not go away when scientists debate rival theories to explain them. Einstein's theory of gravitation replaced Newton's, but apples did not suspend themselves in mid-air pending the outcome. And human beings evolved from apelike ancestors whether they did so by Darwin's proposed mechanism or by some other, yet to be discovered.

This is an example of the circular reasoning that is characteristic of evolutionist writing. Theories are "facts" and data are "facts." Humans evolved from apelike ancestors but we don't actually know the mechanism. He continued:

> Moreover, 'fact' does not mean 'absolute certainty.' The final proofs of logic and mathematics flow deductively from stated premises and achieve certainty only because they are not about the empirical world.

Obviously a “fact” which is only an inference about data could not mean absolute certainty. What’s his point? In logic and philosophy, proofs can flow deductively from stated premises if those premises are true. What are Gould’s stated premises? One is that “human beings evolved from apelike ancestors” even if he can’t say how. If, as Gould wrote, the proofs “achieve certainty only because they are not about the empirical world,” then that would be metaphysics, not science.
Gould continued:

> Evolutionists make no claim for perpetual truth, though creationists often do (and then attack us for a style of argument that they themselves favor). In science, ‘fact’ can only mean ‘confirmed to such a degree that it would be perverse to withhold provisional assent.’ I suppose that apples might start to rise tomorrow, but the possibility does not merit equal time in physics classrooms.

**Non-Perpetual Truth?**

What is truth if it is not perpetual? Non-perpetual truth? His statement, stripped of its sophistry, translates as “Evolutionists make no claim that what they hold and teach is true.” Is that what the school children are taught to believe? No, they are taught to believe it is true. Catholic scholar-priests who teach and publish believe it is true and pervert God’s Revelation to accommodate it. It is easy to understand why Humanists think this is science because they don’t believe in absolutes, just relativism. Who are those perverse people who would withhold provisional assent to “facts” that are only data? Creationists agree with data. By putting the quotation marks around the word fact he indicates he means something other than data. He is saying that persons who withhold provisional assent to evolution are perverse. Obviously that includes Pope Pius XII. His last statement about apples defying the law of gravity is the type of silly stuff professors can say about creationists within the thin air of the elevated Humanist institutions such as Harvard where there is no one to contradict them.

### Don't Bet Your Soul on Non-Perpetual Truth

In summary: Evolutionists begin with the premise that evolution is self-evident, and the observable data are explained best by natural causes, but the explanations must remain theories because they can't be proved or disproved. The interpretations and inferences, they say, are "science" because they involve theories about matters within the physical and life science envelope. It is not asserted that the theory is true and it is not necessary to prove anything because as long as each theory is replaced by another theory it is still science. While the theory may be true it may not always be true. The trick is to always be searching without claiming "perpetual truth" because if they did that they would be no better than creationists. What hubris!

### Religion Does Not Have to be Theistic

Both the evolution model and the creation model use the same observable data to propose different origins. Both are faith-based. On the basis of what its adherents understand to be Divine Revelation from God, a coherent philosophy incorporating origins, morality, ethics, and life's purpose exists under the name of Catholicism. On the basis of what its adherents understand to be the only basis for knowing, namely, human experience and reasoning on that experience, a coherent philosophy incorporating origins, morality, ethics, and life's purpose exists under the name of Humanism. The *Humanist Manifesto I* of 1933 proclaimed the founding of a non-theistic religion and laid out its dogmas in the form of "affirmations of faith" starting with cosmic and biological evolution

# Chapter 16-The Enemy Is Us

On September 10, 1813, American ships under the command of Commodore Oliver Perry engaged a British naval squadron on Lake Erie during the War of 1812. Commodore Perry quickly scrawled a brief report on the back of an envelope and had it sent to U.S. General William Henry Harrison. The first line of his message, *"We have met the enemy and they are ours,"* became one of the most famous naval quotations in U.S. history.

In 1971 there was a popular nationally-syndicated newspaper comic strip called Pogo. Pogo was a 'possum. The animal characters in that strip were known for their seemingly simplistic, but slyly perceptive comments about the state of the world and politics. In 1971 concern about environmental pollution was increasing so on Earth Day the cartoonist drew a scene in which Pogo and his friend Porkypine were walking through a forest littered with trash and Porkypine said "It's hard walking on this stuff." Pogo replied "Yep, son, we have met the enemy and he is us."

That saying, "we have met the enemy and he is us" immediately became an expression that was applied throughout business, industry, government, the military, etc. when persons within organizations were seen to be working at purposes contrary to the best interests of the organization. Nowhere does that saying seem to have more application than to the Catholic Church. In chapter nine I profiled bible scholars from successive generations to illustrate how seminary professors can harm the faith of their students by "scientific method" interpretation of Scripture. The evolutionists Jesuit papal astronomers Fr. Coyne and Br. Consolmagno have been discussed. The relationship of Vatican clerical bureaucrats with the theistic evolution promoting

Templeton Foundation, ex-Dominican anti-creationist Francisco Ayala, and the exclusion of proponents of creationism and intelligent design from the Vatican conference on Biology was mentioned. For an example of how deeply entrenched at the Vatican the evolutionary propaganda is, google "Papal preacher: Intelligent design is faith statement, not science." Read how a Capuchin priest conducted a Lenten meditation for the Pope and Vatican officials based on his opinion that all of those bright scientists such as at the Discovery Institute and the Institute for Creation Research and elsewhere were not practicing science. There is no limit to similar examples of clerical ignorance of modern science.

## Misunderstanding of Science
## Creates Loss of Confidence in Creation Doctrine

Humanists are doing their best to intimidate Catholics from speaking out against their evolutionary dogma upon which their cultural dominance of us depends. Catholics at all levels in the Church sometimes help them. A pastor preaches that "science tells us the 'when' and 'how' but only the Church can tell us 'why'." He has no idea what he's talking about. The absurd performance of Cardinal Pell as Archbishop of Sydney in a debate with Richard Dawkins spread worldwide on YouTube is another excellent example of ignorance and befuddlement in high places. You will cringe from such a display when he describes Adam and Eve as a myth and even the atheist Dawkins understands how that negates basic Catholic doctrine and calls Pell on it. Readers really must view it to believe it and see how shallow a Cardinal's knowledge of our faith can be.
unamsanctamcatholicam.blogspot.com/2012/05/cardinal-pell-richard-dawkins-adam-eve.html

When fiat creation-supporting Catholics question evolutionary science, some sincere Catholics point to the reports that Pope

John Paul II said evolution was "more than a theory." Also, in 2014, Pope Francis, a Jesuit educated at Jesuit institutions in Spain, Ireland and Germany, said ambiguous things about evolution that were popularly understood to be an endorsement of the theory. Those things were said in circumstances that could never be understood as Popes teaching anything official for the Church's belief. An address to the Pontifical Academy of Science has never been considered an exercise of the Pope's Ordinary Magisterium or the forum in which doctrines are presented for the belief of the universal Church, as many encyclicals are. The Pontifical Academy of Science is not one of the Vatican's Congregations. It is an *ad hoc* group whose members change, and there is no requirement that members be Catholic. Indeed, many are not. Yet, those secular media that want to propagate belief in evolution gave those Popes more ink with respect to those personal opinions than they have given to their authoritative teachings, except to pooh-pooh them. Some sincere Catholics absorb such things from the media and "run with them."

**An Illustrative Example**

In earlier chapters, the opinions expressed by Mr. Fitzpatrick in *The Wanderer* newspaper on January 22, 2015, were used as an illustration of how sincere Catholics can get it so wrong because of what they were taught in school. In a follow-up column on March 12, 2015, responding to his critics Mr. Fitzpatrick brought to his defense quotations from communication that two Popes had with participants at conferences of the Pontifical Academy of Science. Mr. Fitzpatrick quotes Pope Francis as having said:

> God is not …a magician, but the Creator who brought everything to life. Evolution in nature is not inconsistent with the notion of creation, because evolution requires the creation of beings that evolve.

Mr. Fitzpatrick continued:

> This is not the first time that a Pope has spoken on the matter. In 1996, also speaking before the Pontifical Academy of Sciences, Pope John Paul II observed that Pius XII's encyclical *Humani Generis* "considered the doctrine of 'evolution' as a serious hypothesis, worthy of a more deeply studied investigation and on a par with the opposite hypothesis...Today more than a half century after this encyclical, new knowledge leads us to recognize in the theory of evolution more than a hypothesis; that there is 'now a significant argument in favor of this theory.'
>
> It must be stressed that neither Pope John Paul II nor Pope Francis mandated that Catholics accept evolution as a matter of faith, certainly not the Godless version propounded by Charles Darwin. Their position is only that Catholics are free to accept that evolution took place, as long as they see it is a process begun and guided by God the Father."

One could easily infer he means that the Popes have taught authoritatively that "Catholics are free to accept that evolution took place, as long as they see it as a process begun and guided by God the Father." But they taught authoritatively no such thing.

It is quite possible that Pope John Paul II did say "that Pius XII's encyclical *Humani Generis* considered the doctrine of 'evolution' as a serious hypothesis, worthy of a more deeply studied investigation and on a par with the opposite hypothesis." The Pope (most likely his speech writer) was simply wrong. No one who read *Humani Generis* completely and consistently could take away that conclusion. The encyclical is an attack on evolution. The English title of that encyclical is *The Human Race: Some False Opinions Which Threaten To Undermine Catholic Doctrine.* The cause for the false opinions was named as evolution in the 5th paragraph of the encyclical. In paragraphs 41and 42 Pius XII prohibited teaching of evolution!

Who can fault Pope John Paul II's speech writer, possibly a Jesuit, for having put into JP II's mouth words which almost the entire Catholic scholar-priest establishment has been saying in misrepresentation of that encyclical since 1950? As Cardinal Ratzinger explained (above), the "certain kind of reduction" of the doctrine of creation has been going on "in the last centuries." In *Aquinas and Evolution* Fr. Chaberek explained that theologians and philosophers don't know natural science well enough to be able to distinguish scientific facts from the materialistic interpretations and dread being called "anti-scientific." He asserted that because their exaggerated esteem or even fear of the "scientific community" makes them unable to question the so-called "scientific consensus" they have adopted the naturalistic paradigm.

### JP II's Opinion Was Outdated in 1996, More So Today

With respect to the sainted Pope, "more than a half century after this encyclical [*Humani Generis*] new knowledge" does not lead "us to recognize in the theory of evolution more than a hypothesis." As has been pointed out throughout this book, the promotion of evolution as a fact has increased to frenzy during that half century, while the discoveries of science have undermined it.

For example, DNA's structure was discovered three years after *Humani Generis* was published. In that same year the greatest paleontological discovery of all time, namely, the Piltdown Man, was proved to be a fake. The term "Punctuated Equilibrium" appeared in the 1970s as a new fig leaf to explain away the lack of Darwin's "missing" transitional fossils. Space probes beginning in the 1970s disproved the Uniformitarian theory. In

1980 a conference of some of the world's leading evolutionary biologists could not describe any mechanism for macroevolution. In 1990 agnostics founded the Intelligent Design Movement.

Between 1997 and 2000, evolutionist scientists Michael Richardson and Stephen Jay Gould published articles in *Anatomy and Embryology, Natural History, and Science* that criticized Ernst Haeckel's famous 19th Century "embryo drawings" for the fakes that they were because 100 years later textbooks were still teaching school children that human embryo development is a recapitulation in the womb of evolution from animal ancestors. In *Natural History*, March 2000, Gould wrote that we should be "astonished and ashamed by the century of mindless recycling that has led to the persistence of these drawings in a large number, if not the majority, of modern textbooks."

Evolutionists, including theistic evolutionist icon Francis Collins who founded Biologos, claimed that 60% of DNA is "junk DNA." They did not know what purpose it served and thought it was from random mutations just like evolutionists thought body parts were vestigial because they didn't know what they were for. In 2012 the Encyclopedia of DNA Elements Project, a years-long research consortium involving over 400 international scientists studying noncoding DNA in the human genome debunked "junk DNA." In 2015 Francis Collins admitted "it was pretty much a case of hubris to imagine that we could dispense with any part of the genome, as if we knew enough to say it wasn't functional. … Most of the genome that we used to think was there for spacer turns out to be doing stuff."

In 2013 *Darwin's Doubt* which exposed the flaws of all of the evolutionary theories and promoted intelligent design became a *NYT* bestseller. Research reported in 2014 demonstrated "Darwin's finches" are de-bunked by 21st-Century analysis of their DNA. Research published in 2014 seems to have falsified a theory regarding the Earth's magnetic field and undermines evolutionary theories about the age of the Earth. Research in 2014 proved that whale pelvic bones are not vestigial but necessary for reproduction. In 2015 geneticists were not sure the "human mutation rate" should be slow, intermediate, or fast and were fudging their research papers to get the result they want. Example after example of the scientific advances since the 19th Century speculations became the standard "facts" in schools and seminaries could be cited.

In summary there is a wide disparity between the confidence in evolution that many lay intellectuals and scholar-priests promote and the doubts and questions in the peer-reviewed science journals to which those in the field submit papers. Papal speech writers and theologians ought to come into the 21st Century.

### Separate Authoritative Teaching from Media Reports

It is important for Catholics to recognize when a Pope is exercising his teaching authority and when he is merely expressing his opinions. As we have seen, Pope Francis is fond of giving interviews on an airplane and giving speeches during which he gives opinions that are simply products of his heterodox Jesuit formation, not exercises of his ordinary Magisterium. A Pope teaches authoritatively when he reiterates the constant teaching of the Church and promulgates it in such a way as to make it clear that it is for the belief of the universal Church as an exercise of his ordinary Magisterium. A speech to the Pontifical Academy of Science does not fulfill those requirements. Neither do a Pope's tweets on Twitter. Ever since 1859 when Darwin published *The Origin of Species*, the authoritative teaching of Popes who addressed evolution/naturalism has been unfavorable to it and said its doctrine undermined true Catholic doctrine. You read some of that authoritative teaching in this book

# Chapter 17-Plan for a "Comeback"

If this writer has not convinced the reader that the Magisterium, recently and clearly through *Humani Generis*, does not endorse, or "permit," belief in evolution—materialistic or theistic—and instead teaches that evolution has polluted theology, philosophy, and understanding of the Bible, he has failed in the purpose for which the book was written. If the reader continues to assert that paragraph 36 permits him to profess theistic evolution, the reader must think he is one of those fully-qualified in science and theology who was "not forbidden" to research and discuss within limits some sort of evolution and who, complying with the rules for that investigation, has submitted his results for the judgment of the Church and received a "thumbs up." If the reader believes he can teach that theistic evolution of any sort is an acceptable belief without the approval to teach it, and which approval was specifically denied by the same encyclical, he must not be one of those Catholics for whom this book was written.

### Have Faith in the Magisterium

But what if the reader is a Catholic who wants to join his mind and heart to what the Church teaches now that he knows the Magisterium's teaching? Does he have the faith in his Church necessary to accept it, even if he still doesn't see it as a problem? At Mass in the Eucharistic Prayer I, the priest invokes "Abraham our father in faith." Reflect on what Abraham chose to believe against all probability. The reader may not think evolution is a problem, but the Magisterium says it is. Can the reader accept that it is? The scientific consensus says evolution is a fact, but the Magisterium does not agree. Can the reader believe the Magisterium is right and the scientific consensus is wrong? The pollution of theology, philosophy, and understanding of the Bible by Catholic belief in evolution has been discussed at length.

Based on the authoritative teaching of Pius XII and his predecessors, this pollution has been a problem, a growing problem. Be sure to read the history of the problem on the website given on page 119. Based on the fairly recent (1995) opinions of Cardinal Ratzinger, quoted in this book, the problem has gotten worse.

Mark Steyn's opinion (quoted earlier) that "changing the culture (the churches, the movies, the TV shows) is more important than changing the politics" should be taken seriously. The Humanist worldview that permeates the media, the institutions in which we educate our children, the language of public discourse, and the very societal air we breathe has been gaining momentum for over 100 years. The election of Trump gives us opportunity but it will be no more than a "speed bump" to the Humanist juggernaut if Catholics go back to sleep. There is no salvation in politics. The solution is Catholic truth that confronts Humanism's religious underpinning, namely, evolutionary cosmology and biology. The present generation of senior persons currently producing theology books, making catechesis policy, and preaching and who have been victims of the generations-long streak of bad theology, philosophy, and biblical interpretation they were taught in schools and seminaries probably can't lead the revival. Parents, younger priests and young intellectuals in the theology and philosophy departments of those newer colleges striving for complete orthodoxy can and must lead it. Otherwise, all the faithful will get from their appointed Church leaders is more "handwringing," as the Humanists "tighten the screws" because Humanism as the faith propelling the public policy can go only in one direction for Christians. Maybe a re-read of the excerpt from Pope Leo XIII's 1884 encyclical *On Naturalism and Freemasonry* found in Chapter 14 will remind the reader of what is in store for the Church when those folks rule.

## Resolve Personal Doubts

So much has been presented about evolutionary cosmology and evolutionary biology in this book that any reader ought to at least have some doubt about what they were taught in school and <u>why</u> it was taught. To have a clear conscience, it is up to the reader to do what is necessary to resolve the doubt. The reader has heard the testimony of the education system, so now he should look at the other side. In this book, the reader was referred to creation-supporting organizations promoting good science and theology. Evolution is a necessity for Humanists, atheists, agnostics, Freemasons, and others who have no sanctifying grace in them. As science, evolution is riddled with zany theories, there are no known mechanisms for how it happened, its predicted evidence from fossils is permanently "missing;" it can't be proved or disproved, and it's nothing but inferences about data that are more logically explained by fiat creation. On the other hand, the reader has Divine Revelation, the Magisterium, and the sanctifying grace to accept the teaching. The alternative is to choose to believe the dogma of the Humanists. Reflect on St. Paul's Letter to the Romans that was quoted in Chapter 12. Does American evolutionary belief deny the glory of God the Creator? Does American culture reflect what St. Paul predicted such a culture would become? You bet it does.

## The Best Defense Is a Good Offense

For decades, Christian leaders and laity have been fighting a rear guard action while trying to preserve our religious and civil rights. Possibly, Catholic passivity is based on the false belief that this is basically a Christian nation and the majority of our countrymen share at least basic values. For that reason, each new outrage is somewhat of a surprise, especially to the clergy. In Chapter 1, the decline of belief was documented. The majority of our countrymen, especially the younger they are, do not share our basic values. In Chapter 14 it was shown that from its earliest days, this country was a zoo full of exotic theological and philosophical speculation. In the 19th Century and first half of the 20th Century, when a college education was a privilege of the few, the leading universities were shaping the minds of the country's future leaders, teachers, parents, and intellectuals toward Humanism. And they still are. It was shown during the discussion of Fr. Bruce Vawter in Chapter 9 that many priests were taught to believe in evolution just as he was as a seminarian in 1946. Hope lies in educating the younger priests and the seminarians even though the seminaries are part of the problem.

## A Revival of Knowledge

This writer proposes to combat Humanism by a Catholic revival of knowledge and faith in the creation doctrines that Cardinal Ratzinger urged, supplemented by education in modern natural science. Today, creationists are mocked and ridiculed for not believing in evolution. Mocked and ridiculed "yes;" debated "no." Make no mistake; ridicule is an exceptionally powerful tool in propaganda and this is what Humanists use. The saying attributed to the law profession regarding how to win a case applies: "When you have the facts, argue the facts. When you lack the facts, abuse your opponent." Humanists abuse creationists because the science is on the side of creation. If Catholics were to be educated on these matters, Catholics would have the gumption to both debate the facts, if Humanists will debate, and ridicule them when they won't. And lay Catholics would have the confidence they now lack to engage in public policy reform. Reject their evolution hoax and their whole edifice built on it falls.

## It's Time for a "Comeback"

A parish bulletin announced that "CCD for K through high school resumes August 30th after the summer break on Sundays 12:00 to 1:00." Is that it? Is that all we have to offer our school children while the public schools have all the time needed to teach them Humanist dogma?

Should clergy and laity consider a new approach to the formation of young Catholics? Because what we have been doing isn't adequate. Is it time for a new approach? At the 1976 Eucharistic Congress Cardinal Karol Wojtyla (JP II) warned:

> We are now standing in the face of the greatest historical confrontation humanity has ever experienced. I do not think the wide circle of the American society, or the wide circle of the Christian community, realize this fully. We are now facing the final confrontation between the Church and the anti-church, between the Gospel and the anti-gospel, between Christ and the antichrist.

On July 24, 2015 Patrick Buchanan wrote:

> "If God does not exist, then everything is permissible." Ivan Karamazov's insight came to mind while watching the video of Deborah Nucatola of Planned Parenthood describe, as she sipped wine and tasted a salad, how she harvests the organs of aborted babies for sale to select customers. "Yesterday was the first time ... people wanted lungs," said Nucatola, "Some people want lower extremities, too, which, that's simple. ... "I'd say a lot of people want liver. ... We've been very good at getting heart, lung, liver, because we know that, so I'm not gonna crush that part, I'm gonna basically crush below, I'm gonna crush above, and I'm gonna see if I can get it all intact." Nucatola is describing how an unborn baby should be killed and cut up to preserve its most valuable organs for sale by its butchers. Welcome to God's Country, 2015.

When the videos of Planned Parenthood selling organs and body parts became public in July 2015, a priest from the pulpit demanded to know, "Where is the outrage?" A pew sitter was seen mumbling softly "Where is the leadership, this is a hierarchical church." I could relate to that. As a Pennsylvania-born teenager I traveled by Greyhound Bus in the South for the first time in 1955. I saw the water fountains, restrooms, and lunch areas marked "colored" for the use of those riding in the back of the bus nearly 90 years after the 14$^{th}$ Amendment said they were citizens and had equal rights. Certainly those citizens were individually outraged and many, many prayers were offered to end it. But the beginning of the end was *organized action*, namely, the Montgomery Bus Boycott of 1955-56. That was led by black pastors Martin Luther King, Jr. and Ralph Abernathy who subsequently formed and led the Southern Christian Leadership Conference which carried forth the organized leadership efforts of black pastors. Black solidarity and taking the moral high ground ended the outrage. In a hierarchical church, fostering Catholic solidarity and preaching and leading to the moral high ground is a clerical responsibility. Christian values are now in the back of the bus. It is an outrage that Christian children are being indoctrinated in the Humanist religion by our taxpayer-funded institutions and the appointed leaders and communication organs of the Catholic Church in America are passive and silent.

### The Time Is Ripe To Challenge Bogus Education

The May 10, 2012 issue of *Nature*, an evolution-supporting technical journal, featured an article "Beware of Creeping Cracks of Bias" which warned that "Evidence is mounting that research is riddled with systematic errors" and "A biased scientific result is no different from a useless one." The problem was most provocatively asserted in a now-famous 2005 paper by John Ioannidis, "Why Most Published Research Findings Are False" (J. P. A. Ioannidis *PLoS Med.* 2, e124; 2005).

The "scientific consensus" is being dissected even by non-scientists. For example, journalist Tom Bethell's 2017 book, *Darwin's House of Cards: A Journalist's Odyssey Through the Darwin Debates* exposes evolution as a 19th-century idea past its prime, propped up by logical fallacies, bogus claims, and "evidence" that is disintegrating under an onslaught of new scientific discoveries. His concise yet wide-ranging tour of the flashpoints of modern evolutionary theory clearly reveals the weaknesses of the theory that rarely are exposed in mainstream literature and education, or the media. Obviously Bethell has a brilliant mind and has been researching this subject for decades because he gives accounts of his interviews with some of the leading evolutionary scientists of the past half century such as Harvard biologists Stephen Jay Gould, E.O. Wilson and Richard Lewontin as well as British paleontologist Colin Patterson. It is a "page turner" full of good science, history and philosophy.

### It is the 4th Quarter

If a football analogy makes sense, the Christian position in America is that Christians are many points behind and it is the fourth quarter. Trump's election is only a "time out" that allows us to regroup. No amount of defense can win the contest. Catholics must go on offense, and that begins with educating the children. As explained in Chapter 2, 48% of those who lose their Faith lose it by age 18. CARA studies indicate they lost Faith by 13. But *parents require real support from their clergy to organize that education.*

### The Coaches

To take the football analogy further, a revival requires coaches, players on the field, and a support group of fans. It is the "coaches" who need to know the most. Physicist-theologian Fr. Victor Warkulwiz's book, *The Doctrines of Genesis 1-11: A Compendium and Defense of Traditional Catholic Theology on*

*Origins,* is a "coach's handbook." Fr. Warkulwiz is one of those unique individuals qualified in science and theology who were "not forbidden" by Pius XII to research and discuss the question of evolution. Fr. Warkulwiz's education and experience prepared him for a work such as this. It has given him professional expertise in physics and theology and an "educational acquaintance" with philosophy, history, and various scientific disciplines. Dr. Warkulwiz, who was once a ballistic missiles analyst at the CIA, spent years working in hi-tech industry, research positions and college teaching before he heard the call to the priesthood. Fr. Victor received a M.Div. from Mount St. Mary's Seminary and an M.A. in theology from Holy Apostles Seminary. He taught courses in literature, mathematics, and physics in the college seminary at Holy Apostles and courses in philosophy and religion at the Franciscan Friars of Mary Immaculate scholasticate. Fr. Victor was ordained in 1991.

**The Plan of the Book**

The purpose of *The Doctrines of Genesis 1-11* is to help restore traditional Catholic theology on origins to its rightful place in the belief of Catholics. The traditional teaching of the Church on Creation, the Fall, and the Great Flood and its aftermath is clearly presented in the form of sixteen doctrines abstracted from the text of *Genesis* 1-11. The doctrines are defended on theological, philosophical, and scientific grounds from assaults made on them from the sectors of biblical criticism and scientism. Accurate, thorough, and readable answers are given on many questions that perplex the modern Catholic. The exposition is kept as non-technical as possible so that the book is accessible to everyone. Not everyone will be able to understand everything that is presented, but every reader will find enough to set his thinking straight and to nourish his Catholic faith.

## Parish Priests Must Be Engines of Catholic Resurgence

This writer does not wish to offend any priest, but he has accepted the testimony of Cardinal Ratzinger that the doctrines of creation have disappeared from priestly formation, and as a result are not part of catechetics or preaching. Fr. Warkulwiz's book could be studied by every priest and could be used as a textbook in seminaries. Priests could learn and teach biblical fiat creation to the people with confidence. Priests have the ear of the laity. They have the special grace of their office. Priests have institutional resources. They control access to parish facilities. Priests can invite knowledgeable creation and natural science speakers to put on programs and encourage the laity to attend them. For example, what would be so hard about having father-son or mother-daughter nights where creation-supporting science videos are watched and discussed? That doesn't cost anything but commitment to evangelize.

Since this education endeavor is for Catholics by Catholics, it is incumbent upon the clergy, after informing themselves, to open the Church property such as parish halls, parish bulletins, and other facilities to the "players on the field." In fact they should invite and lend support to the players. They could ask the players for suggestions of things that could be done, such as starting a parish discussion club on this topic. It is all very simple if there is a will. The play clock is running.

## The Players on the Field

The players are the ones whose concern for souls causes them to practice the spiritual works of mercy by willingness to teach. The players are those laity and clergy who, grasping the importance of the situation, are affiliated in groups that are prepared to lead a discussion club, give public lectures and seminars, who stock their web page with great theology and science, and publish or distribute DVDs, magazines, newsletters, and books. In this regard, the Evangelicals have outstanding capability. The Institute for Creation Research (ICR) founded in 1970 is the "Top of the Top". ICR has a fantastic stable of speakers who are giving seminars at Evangelical churches and pastors' conferences all over the country. ICR's web site (ICR.org) is an encyclopedia of science. ICR's researchers, writers, and lecturers are Ph.D. scientists in numerous disciplines from secular universities. They publish creation-supporting books, DVDs, and a full-color monthly magazine mailed free of charge to all who request it. Obviously ICR receives the kind of financial support needed for such a professional operation from individual Evangelical churches and individuals of all faiths including this writer.

### A Fabulous Seminar

This writer was privileged to witness an example of how any Catholic Diocese (or even a group of parish pastors in a region) could easily accomplish creation and science education on a large scale. In November 2015 I attended a one and a half day seminar at Patrick Henry College that featured ICR scientists. Patrick Henry is an Evangelical institution in a relatively rural area of northwestern Virginia. It was founded by the man who also founded the Home School Legal Defense Association. It opened in 2000 and has about 350 students. The faculty took the initiative to get ICR to come from its Texas HQ. I found out about the seminar because I'm on ICR's mailing list. Attending with me were the headmaster and four teachers from an independent, lay-run Catholic high school. When we walked in for the 3:30 PM Friday start time we were stunned by the size of the crowd, nearly all adults, who filled the floor of a large gymnasium. (Some students were in the bleachers at the back of the audience.) The seminar's theme was "Unlocking the Mysteries of Genesis."

### Real Scientists, Real Education

The following list of presenters illustrates the depth and expertise on display:

- Jason Lisle earned his Ph.D. in astrophysics from the University of Colorado. The topics of his lectures were "Your Origins Matter" and "Astronomy Reveals Creation."
- Tim Clary earned his Ph.D. in geology from Western Michigan University. His lecture was "Geology: The Secrets of the Genesis Flood". (He also led the breakout session that I attended "Oil, Fracking and a Recent Flood.")
- Henry Morris, III, is the son of ICR's founder who started the modern creation science movement with his book, *The Genesis Flood.* Dr. Morris spoke on "Apologetics: Unlocking the Mysteries of Genesis."
- Randy Guliuzza is a Public Engineer and a Medical Doctor. His first topic was "Biology: Made in His Image; Complexity of the Human Body." His talk was devoted to the amazing process of human reproduction. I thought I knew all about that because for many years I was a certified teacher of natural family planning but Dr. Guliuzza's lecture and amazing slides were like a graduate course. And he has a terrific sense of humor that had us splitting our sides. Later he lectured on "4 Biological Facts about Creation."
- Marcus Ross, a Ph.D. in paleontology, is an Associate Professor of Geology at Patrick Henry and the one who organized the seminar. He spoke on "Scripture and Geology: Creation Undone". He explained the amazing parallels between the creation narrative and the Flood narrative.

The whole scene was very professional. Lecturers were on the stage and their slides were on huge video screens, one on each side of the room. A first-rate sound system provided flawless acoustics.

### Interfaith Cooperation in Action

ICR is clearly Evangelical. All of the speakers were very open in confessing their belief in the Bible as God's word and their desire to glorify Him by teaching the marvels of His creation. I had my antenna up to detect anything at all that an orthodox Catholic might find objectionable or counter to anything the Magisterium teaches. There was nothing. On the ride home in the car with the teachers I listened for anything they had noticed but they were all talking about how informative and edifying it was. There is no reason that ICR could not provide the same seminar for a diocese or a group of parishes that had the will to put on something that first-rate. We adults paid just $20. The student rate was $10 and there were additional discounts for groups of 10 or more. All it takes is the will. Any diocese or group of pastors could easily get an even bigger audience especially if it wasn't held way out in the country like the seminar I attended. The Institute for Science and Catholicism would organize it if sufficient support from the clergy was present. See http://www.icr.org/events-host

### Catholic Players

The Kolbe Center for the Study of Creation has presented seminars in North America, Europe, Africa, and Oceana. The Kolbe Center is producing much on a shoestring. A cadre of volunteers support the Center's full-time director, maintain its excellent Facebook and web site (kolbecenter.org), contribute articles, and lecture at seminars. Kolbe is seeking active or retired scientific talent to become volunteer speakers. Kolbe seminars normally cover the doctrines of creation and a presentation of creation-supporting science. These seminars are free and supported by donations of attendees. On a smaller scale, the

Institute for Science and Catholicism which is responsible for this book is doing what it can.

### The Fans

Not every Catholic can be a player but every Catholic can be a supportive fan. At the personal level, one can become informed by following the creation-supporting organizations online, getting their free newsletters by email, and making an occasional donation. Use the "donate" button on scienceandcatholicism.org. Fans can educate their own children by buying or downloading free online resources. Why not order this book to be sent to someone?
For those theistic evolutionists who have not been convinced by anything written here maybe reading *The Shadow of Oz: Theistic Evolution and the Absent God* by Wayne D Rossiter will help. It was published in 2015. Dr. Rossiter left Christianity and became an atheist as a biology student before recovering faith and becoming a biology professor. His book is a brilliant dissection and refutation of all of the leading theistic evolutionary authors in which he shows how they have made God into their own image through faulty philosophy and theology.

Your public library is full of books by evolutionist authors. Ask for your library to buy this book and one or more of the books in Appendix III. All are sold by huge online book sellers.

Fans can ask that the Kolbe Center, ICR, the Discovery Institute or the Institute for Science and Catholicism be invited by their pastors to give a seminar, make a presentation or lead a discussion at the parish. Fans must encourage their priests because there will be no "comeback" unless priests at the parish level organize and lead it. They are the custodians of the property and the pulpits necessary for Catholics to communicate with their fellow Catholics, that is, Catholic to Catholic. The pastor can promote your efforts or ban them by allowing or prohibiting access to parish facilities. If "Father" doesn't support fiat creation but instead holds on to the 19th Century science of theistic evolution, this education won't happen. Children will be left to the Humanists to "educate" and alienate.

The work of the fans is the most important work because only when Catholics regain the supreme confidence in the Bible and the Magisterium can they stop the hemorrhaging of the Catholic youth from the Church and the Humanist domination of America. If you think "this is all very interesting but this doesn't affect me" consider that the culture described in St. Paul's Letter to the Romans (p. 186) is the result of a society intentionally denying the Creator and that is exactly what the culture you live in has become. Who but the Humanists could have demanded transgender restrooms, locker rooms, etc. in schools? What but Humanism motivated Target's management to go transgender? Many commentators on the "transgender issue" have called such actions "insane" and searched for a reason for the apparent loss of common sense. The reason given by St. Paul is that "for though they knew God, they did not honor him as God or give thanks to him, but they became futile in their thinking and their senseless minds were darkened. Claiming to be wise, they became fools." The loss of common sense is a punishment.

In an April 2016 column Patrick Buchanan observed:

> A people's religion, their faith, creates their culture, and their culture creates their civilization. And when faith dies, the culture dies, the civilization dies, and the people begin to die.

Get organized. The Humanists destroying American faith and culture are. Become part of the creation-supporting New Evangelism "family." Today!

# APPENDIX I-What about Galileo?

An enormous number of books have been written in the last 150 years claiming that the Catholic Church is not competent to speak of science. Invariably at some point "the Galileo Case" is trotted out as "proof." This can be intimidating, especially to the young. This appendix will provide the reader with so much detail that he will never have to blink when an antagonist asks "But what about Galileo?" The reader will know more about the Galileo Case than anyone with whom he is likely to have a discussion.

The true story is far different from the popular narrative that has been embedded in our culture by persons who thereby hoped to silence the Church. The appendix begins with an article published by the *New York Times* in 1992 on the occasion of Pope John Paul II's alleged apology because that article is a classic example of the popular narrative that is full of false statements and important omissions. After reading the true story the writer hopes the reader will be as convinced as he is about how wisely and charitably Galileo was judged and treated.

In order to reduce printing costs and make this book more affordable this Appendix is made available to be downloaded from the website of the Institute for Science and Catholicism here.
www.ScienceAndCatholicism.org.
On the home page click on the tab "What about Galileo"

# APPENDIX II-Letter from Ireland

*The letter below was received in 2013 from the Dublin webmaster for the Daylight Origins Society which is the Catholic creation-supporting organization in Great Britain and Ireland that was described in Chapter 11.*

Dear Thomas,

I'll now give you my testimony as to how I got involved with the Daylight Origins Society and how our other Irish members got involved. In the late summer of 2004 I was questioning the existence of God. I remember thinking that when I was a child I could clearly believe that God existed. But in that summer, I was really struggling to believe in Him. I had nothing to go on, nothing to back up the little faith I had left. So I prayed what I thought was to be my last hurrah prayer to God. I prayed with absolute sincerity, asking Him to show me that He really exists. I pleaded with Him... I did not demand to see God, but simply show me strong evidence that God really exists. Within a few days, I typed in 'evolution' on Google. The 1st thing that came up was 'Creation or evolution, does it matter what you believe?' I started to read, and read, and read. I read for over an hour, being intrigued by this pdf- type magazine throwing up questions that clearly showed that there must be an intelligent designer. As I worked as a draughtsman in an engineering office, I could readily understand the processes of conceptualizing something, drawing it, tweaking it, and having it tested before fabrication. For the 1st time I could see God in a new light.

In October of 2004 Dr. Monty White of Answers in Genesis [founded and directed by Ken Ham whose debate with Bill Nye was described in Chapter 11] came to Dun Laoghaire to give a seminar in an Evangelical church. I decided I was going to

attend. Dr. White spoke about Mount St Helens [in the U.S.] and the May 1980 volcanic eruption. He spoke of the subsequent formation of a mini Grand Canyon 1/40th the size as the famous canyon itself. He showed a slide show of Steve Austin, a U.S. creation scientist and his investigative work. Dr. Austin showed how rapid rock formation happened almost before our eyes. Steve Austin spoke of upright petrified trees cutting through supposed "millions of years" of rock strata. So from that seminar, I got the creation gist, and I accepted it as an answer from God to my sincere prayer pleading for strong evidence of His existence. I went around and told everyone who would listen and I found no one interested. It was discouraging but I had my faith back and I was not going to let it die like it almost did ever again.
[See a summary of Dr. Austin's research here
http://creation.com/lessons-from-mount-st-helens ]

In 2005 Ken Ham came to University College, Dublin and I attended. I was now on fire for God and the question I now asked myself was, what about my fellow Catholics? They need to know about this. Another question surfaced in my mind. Maybe I had to make a fundamental choice. Do I stay in the Catholic fold? Or do I join with the Evangelicals? I knew I'd have an easy time selling creationism as an Evangelical. But something told me that the Catholic faith is where I should stay. I owe this hunch primarily to Our Lady of Knock. I had so many happy days messing around the Basilica of Knock as a boy. I remember my parents praying their rosaries.

I next decided to do a Google search on Catholic creationism, and I came across via Wikipedia, the Daylight Origins Society. I found this link and then I had to email some fellow called Donal Foley who had an email for the *Daylight* magazine editor. I sent the editor an email, and I subsequently received my first copy of *Daylight*. I was impressed by the richness of the articles. This was better than the Protestant stuff, which is no easy feat. I called the editor and encouraged him in his work. We kept in contact and before long I started learning to build a website for Daylight.

During these times, I was in a prayer group in Donnybrook called Pure in Heart and in 2006 I began to work as an admin for the ministry there. There were two young leaders to whom I witnessed creation but found little receptivity. However, one of the schools at which these leaders gave a retreat on chastity had a teacher who gave them a DVD called Creation for Catholics. The teacher's name is John Donnelly. I did not waste time hooking up with this guy, and was so thankful to the Lord.

I later went to seminary at Maynooth [which is attached to a secular public college] in 2009 and there was a John Paul II society there offering a seminar competition. Two men threw their hats in, and one was named James Lynch of Donegal. James was a seminarian for the Holy Spirit Fathers, and his talk was completely creationist. I had never met the guy before and thought he must be a Protestant. I kept my distance that night as I was developing a waiting game. After 3 months, and at the close of the academic year, I introduced myself to James and we laughed when I told him I was sure he was a Protestant. James' late uncle was a Catholic priest. James also had Hugh Owen [Director of the Kolbe Center for the Study of Creation] give a seminar in Maynooth that was poorly attended. I would quietly develop my web building knowledge while at Maynooth. I left Maynooth in 2011when the junior formation director asked me to bring my time there to an end.

In 2012 Hugh Owen sent an email looking for venues in Ireland or the UK to give talks on special creation. I laughed to myself thinking of the poor attendance at Maynooth. I'm not getting involved with this I had convinced myself. My crew is too small and I was not certain that we had the ability to pull it off. Hugh later sent me a new Irish contact, Francis McLaughlin. I quickly made contact with Francis, and now I had this surge of energy to arrange a few seminar venues for the Kolbe Center visit to Ireland. In November of 2012, we had a series of reasonably attended seminars. It was a good result for our hard efforts.

That has been the story thus far. This is a long war, and we just have to keep plugging away.

Yours,

Paul

**********************************************

In March 2011, the year Paul was asked to leave, the *Irish Catholic* newspaper reported that Maynooth may be set for closure after the Apostolic Visitation found academic deficiencies and other formation related issues. For example, some of the moral theology taught "is not sufficiently orthodox for future priests."

In August 2016 another homosexual scandal at Maynooth became too much for the Archbishop of Dublin who packed his seminarians off to Rome.

Paul persevered and survived the chaos in Ireland. In 2016 Paul finished his novitiate year and made his first vows with the Incarnate Word, a religious order based in Italy.

# Appendix III-Science and Catholicism

***The Doctrines of Genesis 1-11: A Compendium and Defense of Traditional Catholic Theology on Origins*** (2007) by Rev. Victor P. Warkulwiz.

***Darwin's Doubt: the Explosive Origin of Animal Life and the Case for Intelligent Design*** by Stephen C. Meyer. This is a 2013 *NYT* Bestseller. It is very readable for the ordinary non-scientist.

***Evolution: Still a Theory in Crisis*** by Michael Denton is the 2016 follow-up to his 1985 book that helped ignite the Intelligent Design Movement.

***Darwin's House of Cards: A Journalist's Odyssey through the Darwin Debates*** (2017) by Tom Bethell is a masterpiece of science, history, and philosophy by a non-scientist. This book and the two above are published by the Discovery Institute. At my request my public library bought them and Fr. Chaberek's book below. They had plenty of pro-evolution books by Dawkins, Gould and Mayr. Request your public library to get them.

***Aquinas and Evolution*** (2017) by Michael Chaberek. O.P. explained why St. Thomas's teaching on the origin of species is incompatible with evolutionary theory.

***The Shadow of Oz: Theistic Evolution and the Absent God*** (2015) by Wayne D. Rossiter. A biology professor's arguments against the scientism and philosophy of theistic evolution.

**Kolbe Center for the Study of Creation** (online at kolbecenter.org) has a vast amount of free reading and great books for sale. This is the website for authentic Catholic creation theology and natural science. The Kolbe Center will provide a free seminar to any Catholic group or institution. Follow also on Facebook.

**The Institute for Creation Research** (online at icr.org) is a premier creation-supporting science resource. In addition to so much free information online, ICR sells books and DVDs suitable for all ages. Get *Acts & Facts* free.

**Creation Ministries International** (online at creation.com) Subscribe to CMI's free daily email science articles and get a creation science education day by day. This is a super resource.

**The Discovery Institute's Center for Science and Culture** (online at discovery.org/id/) is a comprehensive resource offering much free information. Combining science and faith they provide many recommended resources. For example, "Discovering Intelligent Design" is a textbook, workbook and DVD combination that presents the best evidence from physics, astronomy, chemistry, biology and related fields that nature is the product of intelligent design rather than blind, unguided processes. The program is designed for home schools, private schools, and personal study. Subscribe and get a free email called Nota Bene. Look at www.evolutionnews.org. Bookmark it.

**The Creation Research Society** is a professional membership organization of scientists and laypersons committed to scientific special creation and a young earth. They publish a great quarterly of scientific importance. creationresearch.org/

**Center for Scientific Creation** (online at creationscience.com/) has published a most comprehensive book on the Flood and the earth's geology called *In the Beginning: Compelling Evidence for Creation and the Flood*. You can read it FREE online.

**Daylight Origins Society**-Creation science in the UK and Ireland www.daylightorigins.com

*************************************************

**Institute for Science and Catholicism** (ISC) is responsible for this book. Copies of *Creation, Evolution, and Catholicism* are mailed free of charge to priests, seminarians, teachers, students and others. Will you support this evangelization effort with a tax-deductible donation to ISC and sent by using the "donate" button at the bottom of our website's home page? www.scienceandcatholicism.org Please "Like" us and follow Institute for Science and Catholicism on Facebook.